Military and Humanitarian Health Ethics

The interdisciplinary book series Military and Humanitarian Health Ethics fosters an academic dialogue between the well-established disciplines of military ethics on the one hand and medical ethics, humanitarian ethics and public health ethics on the other hand. Military and Humanitarian Health Ethics have emerged as a distinct research area in the last years, triggered among other things by the unfortunate realities of armed conflicts and other situations of humanitarian disasters - man-made or natural. The book series focuses on the increasing amount of ethical challenges while providing medical care before, during, and after armed conflicts and other emergencies. By combining practical first-hand experiences from health care providers in the field with the theoretical analysis of academic experts, such as philosophers and legal scholars, the book series provides a unique insight into an emerging field of research of high topical interest. It is the first series in its field and aims at publishing state-of-the-art research, illustrated and enriched by field reports and ground experiences from health care providers working in armed forces or humanitarian organizations.

We welcome proposals for volumes within the broad scope of this interdisciplinary and international book series, especially proposals for books that cover topics of interest for both the military and the humanitarian community, and which try to foster an exchange between the two often separate communities of military and humanitarian health care providers.

Sheena M. Eagan • Daniel Messelken
Editors

Resource Scarcity in Austere Environments

An Ethical Examination of Triage and Medical Rules of Eligibility

Springer

Editors
Sheena M. Eagan
Department of Bioethics & Interdisciplinary Studies, Brody School of Medicine
East Carolina University
Greenville, NC, USA

Daniel Messelken
Zürich Center for Military Medical Ethics, Center for Ethics
University of Zürich
Zürich, Switzerland

ISSN 2524-5465 ISSN 2524-5473 (electronic)
Military and Humanitarian Health Ethics
ISBN 978-3-031-29058-9 ISBN 978-3-031-29059-6 (eBook)
https://doi.org/10.1007/978-3-031-29059-6

This work was supported by Centre of Competence for Military and Disaster Medicine (CH)

This Springer imprint is published by the registered company Springer Nature Switzerland AG
The registered company address is: Gewerbestrasse 11, 6330 Cham, Switzerland

Contents

Editors and Contributors

About the Editors

Sheena M. Eagan is Assistant Professor of Bioethics at the Brody School of Medicine, East Carolina University. She is also Head of the North Carolina Unit of the International Chair of Bioethics/WMA Cooperating Center and Founding President of the American Society of Bioethics and Humanities group for Military, Humanitarian and Disaster Medicine. Dr. Eagan holds a PhD in the Medical Humanities from the University of Texas Medical Branch as well as a Master of Public Health from the Uniformed Services University. Her research and teaching have focused on medical ethics and the history of medicine, with a subspecialized focus on military medicine and Veteran health.

Daniel Messelken is Research Associate at the Center for Ethics at Zurich University and leader of the Zurich Center for Military Medical Ethics. He also serves as Head Ethics Teacher for the Center of Reference for Education on IHL and Ethics of the International Committee of Military Medicine and is member of the Board of Directors of the International Society for Military Ethics in Europe (EuroISME). Dr. Messelken studied Philosophy and Political Science in Leipzig and Paris (1998-2004) and received his PhD in Philosophy from the University of Leipzig in 2010. Besides Military Medical Ethics, his main research fields include Just War Theory, the Morality of Violence, Military Ethics, and Applied Ethics more generally.

Contributors

Martin C. M. Bricknell Conflict, Health and Military Medicine, King's College London, London, UK

Eleanor M. Choi Georgetown University, Washington, DC, USA

Joanne Clifford Defence Ethics Programme, Department of National Defence, Ottawa, ON, Canada

John Dorsch UNC Southeastern Medical Center, Lumberton, NC, USA

Heather Draper Warwick Medical School, University of Warwick, Coventry, UK

Paul C. Eagan Canadian Forces Health Services, Ottawa, ON, Canada
Dalhousie University Faculty of Medicine, Halifax, NS, Canada

Sheena M. Eagan Department of Bioethics & Interdisciplinary Studies, Brody School of Medicine, East Carolina University, Greenville, NC, USA

Dirk Fischer Teaching and Research Unit for Military Medical Ethics, Bundeswehr Medical Academy, Munich, Germany

Frederic Gilbert EthicsLab, Philosophy Program, School of Humanities, University of Tasmania, Hobart, TAS, Australia

Jacob S. Hogue Department of Pediatrics, Madigan Army Medical Center, Tacoma, WA, USA

Simon Horne Centre for Defence Healthcare Engagement, Royal Centre for Defence Medicine, Birmingham, UK

Sarah L. Huffman Director of BioBehavioral Health Research, David Grant Medical Center, Fairfield, CA, USA

Samia Hurst Institut Ethique Histoire Humanités, University of Geneva, Geneva, Switzerland

Robert James Academic Department of Military Emergency Medicine, Birmingham, UK

Erika Ann Jeschke Senior Lead Scientist, Parsons Corporation, Navy Medical Research Unit-Dayton, Dayton, OH, USA

Ivan Kholikov Institute of Legislation and Comparative Law under the Government of the Russian Federation, Moscow, Russia

Bernhard Koch Institute for Theology and Peace, Hamburg, Germany
Department of Philosophy, Goethe University, Frankfurt, Germany
Department of Systematic Theology, Albert Ludwig University, Freiburg, Germany

Stefan Kowitz Multinational Medical Coordination Centre/European Medical Command (MMCC/EMC), Koblenz, Germany

Jeanne A. Krick Department of Pediatrics, Brooke Army Medical Center, San Antonio, TX, USA

Hannah R. Martinez Anchorage, AK, USA

Emily Mayhew Imperial College, London, UK

Daniel Messelken Zürich Center for Military Medical Ethics, Center for Ethics, University of Zürich, Zürich, Switzerland

Tyler R. Reese Department of Family Medicine, Madigan Army Medical Center, Tacoma, WA, USA

Ian Stevens Philosophy & Gender Studies / University of Tasmania, Hobart, TAS, Australia
Department of Neurological Surgery / Oregon Health & Science University, Portland, OR, USA

Matthew A. Studer Division of Cardiology, Department of Pediatrics, Seattle Children's Hospital, Seattle, WA, USA

Elliott M. Weiss Division of Neonatology, Department of Pediatrics, University of Washington, Seattle, WA, USA
Treuman Katz Center for Pediatric Bioethics, Seattle, WA, USA

Stephen N. Woodside United States Military Academy, West Point, NY, USA

Acronyms and Abbreviations

AF SOST	Air Force Special Operation Surgical Teams (United States)
AJMedP	NATO Standard Allied Joint Medical Publication
AMA	American Medical Association
BI	Battle Injury
CAF	Canadian Armed Forces
CASEVAC	Casualty Evacuation
CFHSP	Canadian Forces Health Services Publication
CFJP	Canadian Forces Joint Publication
CIVCAS	Civilian Casualties
CJOC	Canadian Joint Operations Command
COVID-19	Coronavirus Disease 2019
CPR	Cardio-Pulmonary Resuscitation
CT	Computed Tomography
DoD	Department Of Defense (United States)
DoDi	Department Of Defense Instruction (United States)
ED	Emergency Department
EM	Emergency Medicine
FEV	Forced Expiratory Volume
FM	Field Manual
GBA+	Gender-Based Analysis Plus
HADR	Humanitarian Assistance and Disaster Relief
HC	Health Care
HCP	Health Care Personnel
ICRC	International Committee of the Red Cross
ICU	Intensive Care Unit
IHL	International Humanitarian Law
ITU	Intensive Therapy Unit
JESIP	Joint Emergency Services Interoperability Principles
LSCO	Large-Scale Combat Operations

LCP	Lowest Capable Provider
LLE	Life, Limb, Eyesight
LoAC	Law of Armed Conflict
LRS	Limited Resource Situations
MASCAL	Mass Casualty
MEDEVAC	Medical Evacuation (often via air transport)
MedRoE/MROE	Medical Rules of Eligibility
MI	Major Incident
MIMMS	Major Incident Medical Management and Support
MOU	Memorandum of Understanding
MRI	Magnetic Resonance Imaging
MTF	Medical Treatment Facility
NGO	Non-Governmental Organization
NICE	National Institute for Health and Care Excellence (UK)
NHS	National Health Services
ODA	Operational Detachment Alpha
ODP	Operating Department Practitioner
OPP	Operational Planning Process
PAR	Population at Risk
PECC	Patient Evacuation Co-ordination Centre
POC	Protection of Civilians
QALY	Quality-Adjusted Life Year
R2P/RtoP	Responsibility to Protect
RevJWT	Revisionist Just War Theory
SOP	Standard Operation Procedure
SOFA	Sequential Organ Failure Assessment
STANAGs	NATO Standardization Agreements
STEMI	ST Elevation Myocardial Infarction
TMR	Triage by Moral Responsibility
TMT	Terror Management Theory
UHC	Universal Health Care
UNOCHA	United Nations Office for the Coordination of Humanitarian Affairs
VIP	Very Important Person
WHO	World Health Organization

Chapter 1
Introduction and Synopsis

Sheena M. Eagan and Daniel Messelken

1.1 Introduction

Modern medicine consumes vast amounts of resources, ranging from human to technological and financial. In a well-functioning and well-equipped health system, resource allocation considerations rarely impact clinical decision-making as all patients that need care will (eventually) receive it. In light of this, health care providers (HCPs) are often taught to focus on the patient in front of them, driven by a type of patient-centred ethics (of care) that prioritizes the individual person's well-being above the aggregate. Informed by the principle of autonomy and respect for individual self-determination, patients in well-resourced health systems often have the autonomy to make decisions that require high resource utilization with little concern directed to how these choices may impact the care provided to other patients or groups of patients. In fact, it is generally accepted that if resources are sufficient, patients and their providers should be able to make medical decisions unencumbered by considerations related to broader resource allocation (with some caveats related to personal or national health insurance schemes). This means that the average HCP in the best case and in their daily routine spends little time considering how their clinical choices might impact the care available to others—concerns

S. M. Eagan (✉)
Department of Bioethics & Interdisciplinary Studies, Brody School of Medicine, East Carolina University, Greenville, NC, USA
e-mail: eagansh17@ecu.edu

D. Messelken
Zürich Center for Military Medical Ethics, Center for Ethics, University of Zürich, Zürich, Switzerland
e-mail: messelken@ethik.uzh.ch

S. M. Eagan, D. Messelken (eds.), *Resource Scarcity in Austere Environments*, Military and Humanitarian Health Ethics,
https://doi.org/10.1007/978-3-031-29059-6_1

related to how resources are allocated, and who receives care are far removed from clinical interaction and related medical decision-making.

It is only in times of scarcity or austerity—when health systems bend, break, or even collapse—that resource allocation becomes a paramount concern directly impacting clinical decision-making on the individual patient level. Since this systems-level (or aggregate) decision-making is often defined as within the realm of public health, administration, and policy, it can be particularly challenging for HCPs to make resource allocation decisions as it involves a paradigm shift away from traditional patient-centred clinical decision-making to a population-centred approach. However, the COVID-19 pandemic has clearly illustrated the limitations of global health systems, prompting governments and health systems to develop crisis standards of care—forcing HCPs to consider not only the patient in front of them, but also the broader patient population and the general (public) health of their community. COVID-19 triage protocols have aimed to appropriately allocate scarce resources ranging from personnel, to testing capabilities, ventilators, hospital beds, vaccines, and treatments. The impact of resource allocation policies on the individual patient and provide is no longer a hypothetical but rather a daily occurrence.

Looking beyond the crisis of this recent pandemic, modern health care has continued to consume more and more resources making issues related to the fair distribution of resources increasingly relevant. The need for justice in resource allocation and utilization is made even more apparent considering a growing body of literature that highlights systemic bias and structural racism as contributors to health disparities (Bailey et al. 2021; Yearby 2020). Inequality and social injustice as a threat for (global) health has also been recognized as part of the broader conversation related to social determinants of health (Marmot 2015). It appears that even well-resourced health systems are only well-resourced for some segments of the population, while other groups of patients struggle with on-going austerity (Bailey et al. 2021; Yearby 2020). Accordingly, greater attention must be paid to issues of resource allocation. Given the somewhat recent focus on these considerations in most well-resourced countries, little attention has been paid to this topic within the field of classic bioethics. The discussion of ethical issues related to resource scarcity have been limited to the context of disaster bioethics, and military medical ethics.

This book offers insights, reflections, and analysis from subject matter experts with experience in the fields of disaster, military, and humanitarian medicine. As the following chapters will show, austerity and resource scarcity are common in, if not characteristic of, the environments where military, disaster, and humanitarian medicine are practiced. In this setting, (local) health systems may be non-existent, overwhelmed by disaster (both natural and man-made), or on the verge of collapse. Beyond health system infrastructure itself, medical care is further challenged by the temporal nature of this care, as well as security threats and military mission constraints. Considering these realities, the practitioners and scholars in these fields have long grappled with and discussed the ethical issues related to resource allocation in times of scarcity. The remainder of this introduction will offer a brief introduction to key concepts and contexts explored throughout the book.

1.2 Disaster & Crisis Focused Interventions: A Spectrum of Activities United by Scarcity

Humanitarian Assistance can be defined as "aid to an affected population that seeks, as its primary purpose, to save lives and alleviate suffering of a crisis-affected population" (quoted in Eagan 2018). The types of health care provided in the humanitarian sphere varies greatly depending on the context, location, and specific needs of the impacted community—however, the focus on crisis highlights the austerity of the settings in which this type of medicine is practiced. Humanitarian assistance includes everything from trauma care following an earthquake or tsunami, to epidemic control, refugee assistance, or civilian aid in areas of conflict, and even health system stabilization. Military and humanitarian health care takes place along a vast spectrum; each of these contexts brings differing levels of resource scarcity that require unique triage schemes or allocation plans to maximize benefit for the broader crisis-effected population. In this way, the humanitarian sphere stands at the intersection of clinical ethics and public health ethics by attempting to care for individual patients in a way that is just and sustainable for the broader population. The goals of the various actors in this space also impact how resources are allocated. Specifically, the objectives of Humanitarian Organizations, Non-Governmental Organizations, Governments, and Militaries often differ from each other, while also differing across the spectrum of Humanitarian Assistance and Disaster Relief activities.

Humanitarian assistance activities are differentiated as being either proactive or reactive by separating disaster relief from other forms of aid. Despite this differentiation, Humanitarian Assistance and Disaster Relief are often discussed together under the acronym HADR. The spectrum of these activities can be further divided into three general categories, all of which may involve resource allocation decisions in austere settings. These categories are as follows:

1. Direct Assistance: This type of assistance is characterized as the face-to-face distribution of goods and services to an affected population. Direct assistance includes individual patient care as well as public health interventions; it ranges from diagnosis, to intervention, including surgery, dental care, health education, the distribution of necessities (ranging from medication to hygiene materials, water, food, etc.) and even veterinary care for livestock.
2. Indirect Assistance: This type of assistance is characterized as activities that are at least one step removed from an affected population. Indirect assistance does not include direct patient care but could include the transportation of medicine/other supplies, and health care personnel.
3. Infrastructure Support: This type of assistance is characterized as activities that support or facilitate relief but are not necessarily of benefit to only the affected population. Infrastructure support includes the building of clinics, roads, bridges, or air strips (often meant for incoming supplies/personnel), as well as logistical support in managing the resources necessary for relief to an affected population/area.

Each type of assistance (described above) involves the import of goods and services to aid a population in need, understood as being in crisis. This crisis-characterization generally involves a health system that has collapsed, is overwhelmed by current need, or not developed enough to handle the affected population. Put another way, humanitarian intervention is necessitated by austerity and scarcity.

At this point, it is important to note that there is a lack definitional clarity surrounding what a "disaster" or "crisis" is. These terms are hard to define and there exists a vast difference in the concepts and definitions used in identifying and forecasting disasters, as well as a lack of clarity concerning when a disaster or crisis is over (O'Mathúna 2018). Especially when focusing on collapsed or under-developed health system, it is hard to know when the crisis has ended and what role outside intervention should play in health system stabilization or (re)establishment. This lack of clarity has practical implications ranging from how and when organizations respond, to when they leave, and what health-related issues will be the focus of their intervention O'Mathúna 2018). Specifically, these types of considerations can impact which patients are eligible to receive assistance. Since disasters are often seen as events that arise in crises that are sudden or unexpected nature, chronic infrastructure or health issues might be seen as beyond the scope of HADR activities. As an example, it might be acceptable for HADR actors who are responding to an earthquake to set broken limbs, but not to address chronic health issues, such as a lack of nutrition that contributes to the broader poor health of the crisis-affected population.

Recognizing that their role is often short-term and meant to offer temporary aid in crisis, practitioners of humanitarian and military health care must often determine strict categories related to eligibility of care, prioritizing those that stand to benefit most and are in line with the goals of that specific mission. This reality stands in stark comparison to most clinically based ethical approaches that would not permit the categorical denial of care to whole patient populations. However, in the humanitarian setting they often do not have the diagnostic equipment, expertise, or resources to treat everyone or everything. Beyond that, the crisis-focused model of humanitarian intervention is often not well-equipped to rebuild or develop local health systems capacities. Instead, decision-making leans toward a utilitarian model, attempting to maximize benefit with limited resources. Beyond that, the temporary nature of this assistance involves the import of personnel and supplies that are not left behind at the end of the mission, challenging the ability to follow-up and often failing to bolster local health systems. In recognition of this, the practitioners and thinkers included in this edited volume offer discussion that often differs from more clinically centred bioethics and traditional public health ethics (often focused/developed in resource-rich nations).

1.3 Humanitarian & Military Actors: Differing Goals & Intention

Another important consideration in the austere setting, is the goal of the specific entity providing the aid in question. Of course, goals will vary from mission to mission—the objectives of assistance in epidemic control will vary widely from those in earthquake (and other disaster) relief/response. While some long-term humanitarian missions might focus on sustainability and development of local health systems, most are more specifically focused on short-term crisis-focused intervention and response. Additionally, each type of humanitarian assistance (direct/indirect/infrastructure) often involves different actors with different skillsets. Generally, direct assistance has been understood as reserved for humanitarian actors, such as NGO's and other humanitarian organizations. Within the humanitarian sphere, military involvement has often been limited to security and logistic support (indirect assistance and infrastructure support). However, militaries are increasingly engaged in humanitarian assistance both in and out of conflict-zones, even providing direct (face-to-face) assistance. This shift has been driven by both a growing need for HADR activities in unstable areas (often active conflict zones), and the shifting goals of militaries around the world. In fact, most developed nations have gradually made HADR an increasingly significant part of military operations, recognizing the strategic and diplomatic value of this work. However, the goals and priorities of militaries remain often different from non-governmental organizations and other more traditional humanitarian agencies.

According to international humanitarian law, humanitarian assistance must be provided in accordance with the core principles of humanity, neutrality, and impartiality. According to the Oslo Guidelines, promulgated by the United Nations Office for the Coordination of Humanitarian Affairs (UNOCHA) in 1994 and revised in 2007, assistance must be based on actual (medical/health-related) needs and delivered by actors that have no political interest or stake in the situation (United Nations 2007). This doctrine is clear that the involvement of foreign military and civil defense assets should be limited to a last resort, where there is no comparable civilian (non-military) alternative. The 2015 Ebola virus disease epidemic in West Africa offers an example of this, during which humanitarian organizations called upon governments and militaries to join in assistance. This call was informed by recognition that militaries were uniquely positioned to offer not only needed logistical/infrastructure support but also uniquely equipped to rapidly deploy personnel, and other resources (Messelken and Winkler 2018). Throughout this global health crisis, military-actors provided all levels of assistance (direct/indirect assistance/infrastructure support). This reality was also true of the COVID-19 pandemic when militaries around the world offered all forms of assistance both domestically and abroad.

Traditional concepts surrounding resource allocation are further complicated by military goals in settings of austerity—militaries (and the governments sending them) often have differing priorities from their humanitarian and non-governmental counterparts and work with shorter and defined timeframes. While all are motivated to offer aid to populations in need (on some level), the prioritization of this goal and its place within the broader mission varies greatly. Generally, militaries prioritize their mission above all else (missions range from purely humanitarian, to offensive military missions in active conflict zone, or the stabilization of an area in low intensity conflict, among others). Beyond that, they often perceive obligations to their own service-members above their obligations to others. Their duties to service-members extend beyond relational obligations and are grounded in a need to maintain force readiness to ensure the ability to accomplish the broader military mission. In light of this reality, militaries have developed policies to determine not only how resources should be allocated, but which patients will be eligible for care. While these policies differ internationally, they generally challenge the paradigm of traditional medical ethics (be they clinically oriented or humanitarian-grounded) by necessitating distinction between patient groups based on affiliation rather than medical need. Constraints related to security, military necessity and patient identity inform these policies and often determine an individual's eligibility for care. These policies are often referred to as Medical Rules of Eligibility and represent a particularly nuanced area of military medical ethics. The field of military medical ethics has a vast body of literature dedicated to resource allocation in scarcity, that is discussed throughout this edited volume. Despite differences in priorities and intent, it is resource scarcity that most obviously connects military, disaster, and humanitarian medicine. In both of these austere settings, resources scarcity is the rule not the exception.

The difference between military and humanitarian intent are particularly apparent in major incidents or mass casualty events in conflict zones. It is within these settings that we see triage that excludes the most severely injured and begins to differentiate patients based on non-medical factors. Triage is perhaps the clearest example of how military medical ethics can inform broader medical ethics in times of austerity; triage was developed on the battlefield and has now proliferated all forms of medical care (from the doctor's office to the Emergency Department, and beyond. In the next section of this introduction, we offer a brief overview of triage.

1.4 Origin and Definition of Triage

The word "triage" is derived from the French verb "trier", which means to sort something, and which was originally used in agriculture (sorting according to quality). It made its way into the medical context and its current medical meaning by way of the Napoleonic battlefield. Napoleon's surgeon in chief, Dominique-Jean Larrey, is most often accredited with being the inventor of the modern idea of triage, although it seems that there is no record of him using the word "triage". Larrey summarizes his approach as follows:

> Those who are dangerously wounded should receive the first attention, without regard to rank or distinction. They who are injured in a less degree may wait until their brethren in arms, who are badly mutilated, have been operated on and dressed, otherwise the latter would not survive many hours; rarely, until the succeeding day. (quoted from Iserson and Moskop 2007, 277)

Larrey insists on prioritizing patients according to their medical need only with the aim of saving as many wounded as possible. As a result, the less severely wounded will have to wait and this also implies that they may not be able to return to the fighting as quickly as if they were treated. Non-medical criteria like rank are explicitly excluded by Larrey. Consequently, one can say that Larrey's reasoning is more that of a doctor who wants to efficiently use his limited medical capacities and much less that of a military service member who might rather aim "To Conserve Fighting Strength" (motto displayed on the US Medical Field Service School's distinctive unit insignia). Larrey's description of (military) triage shows that even if it is a military concept by birth, it was medical criteria that originally that guided the physicians' decision-making regarding the order of treatment.

Similarly, the Encyclopedia *of Bioethics* defines triage as "the *medical* assessment of patients to establish their priority for treatment." (Winslow 2003, 2520, emphasis added) In his book on *Triage and Justice*, Winslow specifies that situations of triage occur under three conditions: there is scarcity of some *life-saving* medical resources; this scarcity must be *obvious*; and the shortage must be *expected to continue* for some (indefinite) period of time (cf. Winslow 1982, 36f). Another helpful definition reads as follows:

> The term 'triage' refers to the procedures clinicians use to prioritize prospective patients. In the background is the unhappy truth that, when vital resources are limited, some will not get what they need, at least not right away. (Kipnis 2002, 1)

Triage can become necessary in different settings and circumstances, both in military and civilian contexts. Their common denominator is that the number of people needing medical care (greatly) overwhelms the immediately available medical capacity. It is thus different from resource allocation insofar as the timeframe is shorter and the decision is more about the outcome for a group than for individuals.

One can speak of a continuum of triage scenarios "based on the ratio of resources to the number of patients who must be evaluated and treated simultaneously." (Iserson and Moskop 2007, 278) Depending on when one starts to speak of "overwhelming", triage may be an issue in the emergency department (ED) and in intensive care units (ICU) or one may delay referring to triage in incidents with multiple casualties (explosions, train accident, etc.), in military (battlefield) triage, or in disaster (mass casualty) triage. In the military context, one might for example make further distinction between mass casualty events (MASSCAL) or a "simple" overwhelming situation in a small unit, when one medic is confronted with three severely wounded.

1.4.1 Triage Categories

In practice, triage is usually implemented by sorting patients into four groups, sometimes amended with a fifth group of those who are already dead. Winslow, in the encyclopedia of bioethics, lists the following categories:

> 1. Minimal. Those whose injuries are slight and require little or no professional care.
> 2. Immediate. Those whose injuries, such as airway obstruction or hemorrhaging, require immediate medical treatment for survival.
> 3. Delayed. Those whose injuries, such as burns or closed fractures of bones, require significant professional attention that can be delayed for some period of time without significant increase in the likelihood of death or disability.
> 4. Expectant. Those whose injuries are so extensive that there is little or no hope of survival, given the available medical resources. (Winslow 2003, 2521).

A similar categorization is implemented by NATO that describes the patient categories in the following way:

> T1/Immediate/Red: Patients with life threatening injuries but a high chance of survival when treated.
> T2/Urgent/Yellow: Patients with severe injuries but not in immediate life threatening conditions.
> T3/Minimal/Green: Patients with minor injuries able to care for themselves. Dead/White on Black: Patients declared dead by a medical professional or with non-survivable injures and no vital signs.
> T4/Expectant/Blue on White: Patients expected to die. Only to be used in mass casualty incidents and when authorized by the Commander or MEDAD/MEDDIR. (NATO 2019, C2).

To demonstrate how these triage categories work and to illustrate how and under what conditions ethical issues are prone to emerge, let us look at three scenarios. The scenarios mainly differ regarding the ratio of available resources compared to the number of potential patients.

The first scenario in Fig. 1.1 shows a situation where the wounded have been sorted into three triage categories and the additional category "dead". The resources in the scenario are then sufficient to treat those falling into the categories T1 and T2, whereas the minimally injured T3 are not treated. Given the circumstances during which triage is applied, such a scenario is acceptable as everybody with a major injury can get treatment. Ethically, thus, it does not lead to major challenges that are due to the distribution of resources.

In the second scenario (Fig. 1.2), resources are no longer sufficient to treat all potential patients in the T1 and T2 categories – those who *could* survive if they received adequate (resource-intensive) treatments. This is when important ethical issues arise as it means that resources must be allocated to some patients and will not suffice for all patients. Some patients will remain without (adequate/lifesaving) treatment despite the fact that their medical status is not beyond hope.

In such situations, the additional triage category "T4 expectant" is introduced to cover persons whose injuries are too serious to be treated with the limited resources available (Fig. 1.3). Saving (or trying to save) these persons would mean to

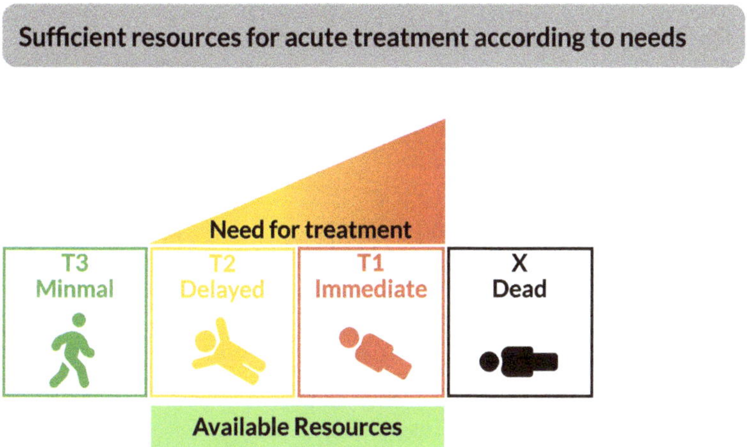

Fig. 1.1 Triage scenario 1

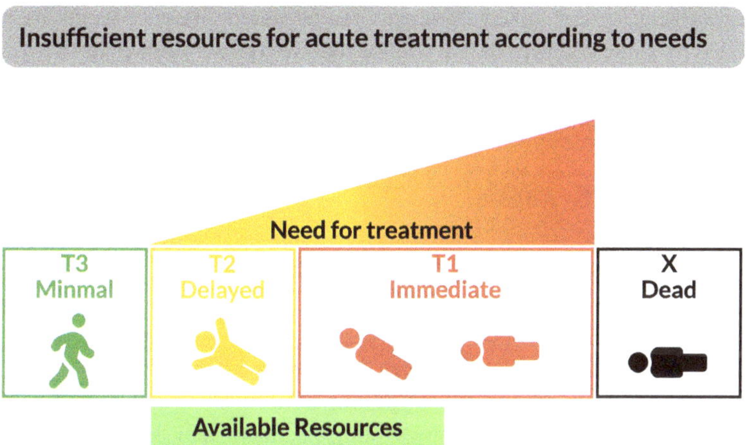

Fig. 1.2 Triage scenario 2

concentrate too many resources per person in the situation. During triage in situations like Fig. 1.3, patients in T4 do not receive life-saving treatment but only sedation etc. to alleviate their suffering before or during dying. Ethically, such a situation obviously is more difficult to handle as it makes choices between patients necessary, and the selection needs to be justified. Such scenarios are sometimes referred to as "tragic choices" (cf. Hunt et al. 2012).

Generally, HCP have an ethical duty to deliver the best possible care under the circumstances, and to do so in an equitable manner. There are however different approaches of how resources should be distributed (see e.g. Persad et al. 2009). Broadly speaking, one can distinguish egalitarian ("treat everybody fairly") and utilitarian ("save the biggest number") approaches to justify triage (criteria) to

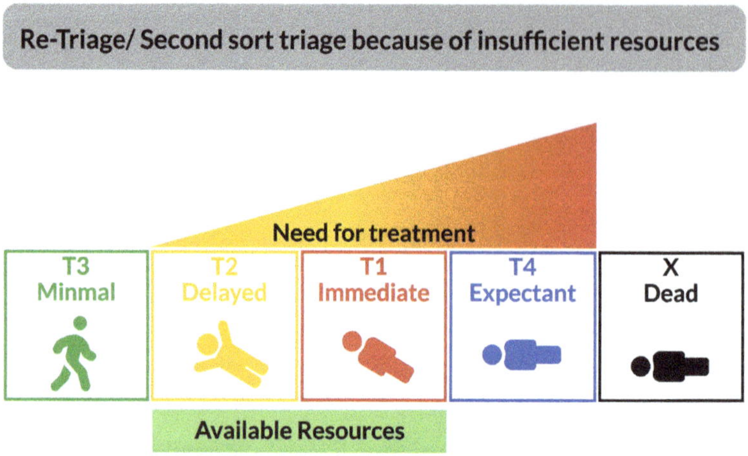

Fig. 1.3 Triage scenario 3

which contractarian and mixed arguments can be added. These concepts are further elaborated by some chapters in this volume, and we refrain from elaborating on this in the introduction.

1.5 Outline & Synopsis

This edited book offers a balance of theoretical analysis and case studies; with chapters exploring real-world examples and practical challenges, as well as international law and military policy. This volume is a product of the ongoing work of the Center of Reference for Education on IHL and Ethics of the International Committee on Military Medicine (ICMM). The contributing authors are drawn from the international faculty and include philosophers, ethicists, healthcare providers, policymakers, and commanders—each with unique experience studying and working in the field of topic.

1.5.1 Section I

The **first section is descriptive** and focuses on the real-life challenges of triage and medical eligibility. Drawing on case studies and policy analysis from a variety of national militaries, this section presents diverse perspectives ranging from commander and policy maker to practitioner in the field. Each case offers insight into how various entities have operationalized resource allocation decisions in policy and practice. While each chapter contributes unique examples of policy and practice, the section as a whole also provides opportunity for critical comparative analysis.

1.5.1.1 'Gate-Keeping Access to Deployed Military Health Services' – Perspectives of Military Medical Leaders—Bricknell and Kowitz

The first chapter provides an overview of the ethical issues related to triage and medical rules of eligibility/entitlement from the perspective of the military commander. Military medical ethics often focuses on the provider, without paying adequate attention to command. Bricknell and Kowitz explore how the requirements of International Humanitarian Law are translated into guidelines of triage, medical eligibility, and entitlement to care. The authors stress the importance of military medical ethics education and explain the role of medical ethics during mission planning. Drawing on a hypothetical scenario, the chapter considers legal, ethical, and clinical principles that might determine possible resource allocation policies.

1.5.1.2 Reconsidering Triage: Medical, Ethical and Historical Perspectives on Planning for Mass Casualty Events in Military and Civilian Settings—Horn, James, Draper, and Mayhew

This contribution focuses on resource allocation during mass casualty events. The authors draw on historical precedent to argue the necessity of planning and training that considers the ethical implications of this type of care. Specifically, the chapter proposes planning that takes a preventive approach to ethics, by preparing providers for established and recurring ethical dilemmas. Among their recommendations, Horn et al. call for the need to deviate from the 'gold-standard', a focus on ensuring patient flow, self-care, remote monitoring, and implementation of the principle that care should be given by the lowest capable (and available) provider.

1.5.1.3 Medical Rules of Eligibility: A Comparative Analysis—Clifford, Eagan, and Eagan

This next chapter provides analysis of the core policies established to define medical eligibility of populations (and individuals) within NATO, the Canadian Armed Forces (CAF), and the United States Department of Defense (DoD). Within the military, the formal doctrine determining who is eligible for care and the level of medical care that they will receive are known as medical rules of eligibility or MROE; however, these rules and how they are developed varies. This chapter offers a comparative analysis of established policies surrounding MROE medical doctrine, providing an overview of how medical eligibility decisions are formalized into policy across military contexts. The analysis reveals that formal doctrine not only varies based on organization/entity but undermines concepts of medical necessity and universality by codifying group-based preferences grounded in the concepts of utility and military necessity.

1.5.1.4 Ethical and Legal Basis for the Standards of Triage Used in the Russian Military Medical Service—Kholikov

The final chapter in the first section adds to the preceding by offering insights from another national perspective—the Russian Military Service. Kholikov provides an overview of how the obligations required under International Humanitarian Law have been historically operationalized into policy and practice within the Russian Federation. Focusing on the standards of triage used in the Russian military medical service, the author argues for the application of objective criteria to assess the severity of the injury throughout every medical evacuation. Arguing that ongoing assessment is a cornerstone of the triage, the author highlights unique factors informing the Russian approach.

1.5.2 Section II

Building on the descriptive nature of the first section, the **second section** of the book takes a more theoretical and/or normative perspective. Authors analyze several ethical challenges posed in situations of resource scarcity and propose approaches to better deal with them. The ethical issues and questions highlighted in this section move discussion beyond the world of military and humanitarian medicine, to highlight the topics' applicability to the general medical practice and policy.

1.5.2.1 The Phenomenon of Allocation. Military Pathways in the Light of Biomedical Ethical Principles—Fischer

The first chapter of the section focuses on the ways that military practice and policy can inform civilian disaster planning. Fischer discusses the relevance of clinical ethics considerations and the patient-provider relationship to discussions of resource allocation in austerity. Focusing specifically the four well-known principles of biomedical ethics (autonomy, non-maleficence, beneficence, justice), the author argues that concepts and principles from military medical ethics are more ethically appropriate in this context and should be translated in everyday civilian policy/planning.

1.5.2.2 Battlefield Triage and Resource Allocation During a Pandemic: What Can We Learn from the Past and How Must We Adapt for the Future?—Krick, Hogue, Studer, Reese, and Weiss

In this chapter, the authors focus on how the lessons learned in military (battlefield) triage can be applied to times of resource scarcity in civilian medicine, such as a public health emergency. Recognizing the differences between the civilian and military context, the authors explore ethical principles and priorities that have guided decision-making in battlefield triage. The chapter focuses on the COVID-19 pandemic, demonstrating how military triage models and decision-making may be utilized during public health crises.

1.5.2.3 Medical Triage by Moral Responsibility—Woodside

The next chapter in this section argues that in times of austerity ethically sound triage decisions can (and should) be based on factors beyond the patient's medical condition. Woodside proposes the concept "triage by moral responsibility," whereby triage decisions are made based on an individual's moral responsibility for their status or "medical predicament." The author illustrates his concept through case studies and argues its broad applicability from military to disaster medicine and pandemic response.

1.5.2.4 Medical Rules of Eligibility – Can Preferential Medical Treatment Provisions Be Ethically Justified?—Messelken

Ethical justifications of medical rules of eligibility are examined in this chapter. It is mainly concerned with the question whether non-medical criteria can legitimately play a role in defining medical rules of eligibility. The chapter explains how MROE are an integral (and necessary) part of the military trauma care system and illustrates this with an example taken from the ISAF experience in Afghanistan. Ethical arguments in favor and against MROE are presented and discussed. Messelken concludes that there are some ethical arguments based on which non-medical criteria in MROE can be defended, but at the same time argues for caution as upholding medical neutrality and impartial medical care is important, too.

1.5.2.5 Fairness in Military Care: Might a Hybrid Concept of Equity Be the Answer? —Gilbert, Stevens, and Hurst

This chapter examines the concept of equity at both the descriptive and normative levels. Gilbert, Stevens, and Hurst argue that while this concept is critical to the distribution of scarce resources, it lacks consensus. Analysis works from the assumption that multiple interpretations of equity exist, arguing that it is not clear which concept is best applied to medical care. This contribution offers both a theoretical overview and empirical data from practitioners in the field. Four major philosophical definitions are explored (Egalitarianism, Prioritarianism, Desertism, and Sufficientism) before a hybrid position of equity is offered. The proposed position is informed by the practice of micro-allocation that is already used by physicians and is presented as both theoretically robust and applicable in practice.

1.5.2.6 Cicero and the Problem of Triage. Why There Is No Moral Algorithm in Distributing Scarce Resources—Koch

This chapter applies an ethical model drawn from the work of Cicero, connecting contemporary issues to ancient philosophical tradition, and focusing on the moral agency of those involved. The author proposes a model of applied ethics that distinguishes between four different moral roles and suggests that doctors (or those who are responsible to select among potential patients) act as moral judges. Reframing the discussion around the moral agency of those who triage, the author explores epistemic questions related to who should triage and how their decisions should be made.

1.5.2.7 Facing Death: An Ethical Exploration of Thanatophobia in Combat Casualty Care—Jeschke, Martinez, Choi, Dorsch, and Huffman

To close the volume, Jeschke and colleagues offer a somewhat different perspective by looking to the future of large-scale combat situations. While the majority of this book focuses on lessons-learned and reflection, the final chapter offers a turn to the austerity of future armed conflict. The authors argue that military medical care was built on the paradigm of *saving lives* and has been enormously successful throughout the conflicts of the past two decades, with an unprecedented survival rate that exceeds 90%. Looking to potential mass casualty events, this chapter argues that systems will be overwhelmed in new ways that challenge the established paradigm and will force us to confront significantly more death and dying. However, medical decision-making has not been explicitly addressed death and dying in military medical training, research, or policy. Relying on a body of literature known as terror management theory and ethnographic data, the authors argue that it is important to explicitly cover topics of death and dying in mass casualty and triage training.

Note from the Editors

This volume was finalized during the first year of the war in Ukraine (2022/2023). Although the topic of this volume holds relevance to the ethical issues faced in Ukraine, the book does not specifically address this ongoing conflict. There are two main reasons for this. First, the work included in this book was started and sometimes completed prior to the Russian invasion. Second and more importantly, an analysis of ongoing conflicts is challenging and beyond the scope of this book. Although many of the issues and insights from this work may be applicable to the war in Ukraine or other ongoing conflict, we do not think that this is the appropriate time or place to address them. New information is constantly emerging, meaning that any analysis would be based on incomplete information that would likely be outdated by

(continued)

the time of publication. Additionally, this book aims to offer broader insights/reflections that are not bound to specific conflicts or crises. As part of a broader series on military and humanitarian health ethics it brings together experts and practitioners in the fields of humanitarian/disaster health, military health care, philosophy, ethics, and other disciplines to provide analysis on a variety of related topics ranging from case studies and first-hand experiences to policy and philosophical analysis.

Acknowledgments Our sincere thanks go to the contributors of this volume who conscientiously prepared their chapters and revised them during the editing process. It is obvious that without their efforts, this book could not have been edited.

Financial support for the work on this volume was granted by the Centre of Competence for Military and Disaster Medicine of the Swiss Armed Forces.

We would also like to thank the anonymous referees for their constructive comments on an earlier version of the manuscript, as well as the team at Springer for their support throughout the conception and production of this volume.

References

Bailey, Zinci D., Justin M. Feldman, and Mary T. Bassett. 2021. How structural racism works—Racist policies as a root cause of U.S. racial health inequities. *New England Journal of Medicine* 384 (8): 768–773.

Eagan, Sheena M. 2018. Global Health diplomacy and humanitarian assistance: Understanding the intentional divide between military and non-military actors. *BMJ Military Health* 165: 244–247. https://doi.org/10.1136/jramc-2018-001030.

Hunt, Matthew R., Christina Sinding, and Lisa Schwartz. 2012. Tragic choices in humanitarian health work. *The Journal of Clinical Ethics* 23: 338–344.

Iserson, Kenneth V., and John C. Moskop. 2007. Triage in medicine, part I: Concept, history, and types. *Annals of Emergency Medicine* 49: 275–281.

Messelken, Daniel, and David Winkler. 2018. *Ethical challenges for military health care personnel: Dealing with epidemics*. New York: Routledge.

Marmot, Michael G. 2015. *The health gap: The challenge of an unequal world*. New York: Bloomsbury Press.

NATO. 2019. Allied joint doctrine for medical support. AJP-4.10(C). NATO Standardization Office.

O'Mathúna, Donal. 2018. *Vilius Dranseika, Bert Gordijn. Disasters: Core concepts and ethical theories*. Springer.

Persad, Govind, Alan Wertheimer, and Ezekiel J. Emanuel. 2009. Principles for allocation of scarce medical interventions. *The Lancet* 373: 423–431.

United Nations. 2007. *Guidelines on the use of foreign military and civil defence assets in disaster relief*. https://www.unocha.org/sites/unocha/files/OSLO%20Guidelines%20Rev%201.1%20-%20Nov%2007.pdf Last Accessed on 22 Sept 2022.

Winslow, Gerald R. 1982. *Triage and justice*. Berkeley: University of California Press.

———. 2003. Triage. *Encyclopedia of Bioethics* 5: 2520–2523.

Yearby, R. 2020. Structural racism and health disparities: Reconfiguring the social determinants of health framework to include the root cause. *Journal of Law, Medicine, and Ethics* 48: 518–526.

Sheena M. Eagan is an Assistant Professor with the Department of Bioethics and Interdisciplinary Studies in the Brody School of Medicine at East Carolina University. Dr. Eagan holds a PhD in the medical humanities from the Institute for the Medical Humanities at the University of Texas Medical Branch, as well as a Master of Public Health from the Uniformed Services University. Sheena is the creator and founding president of the American Society of Bioethics and Humanities group for Military, Humanitarian and Disaster Medicine. She has published articles in peer-reviewed journals, military-specific journals, and contributed to edited books on a variety of topics in military medicine. She has also given talks and lectures on her work in North America, Asia, Europe, and the Middle East. Her research and teaching have focused on medical ethics and the history of medicine, with a subspecialized focus on military medicine.

Daniel Messelken is a research associate at the Center for Ethics at Zurich University and leader of the Zurich Center for Military Medical Ethics (www.cmme.uzh.ch). He also serves as Head Ethics Teacher for the Center of Reference for Education on IHL and Ethics of the International Committee of Military Medicine and is member of the Board of Directors of the International Society for Military Ethics in Europe (EuroISME). Dr. Messelken studied Philosophy and Political Science in Leipzig and Paris (1998–2004) and received his PhD in philosophy from the University of Leipzig in 2010. Besides Military Medical Ethics, his main research fields include Just War Theory, the Morality of Violence, Military Ethics, and Applied Ethics more generally.

Chapter 2
'Gate-Keeping Access to Deployed Military Health Services' – Perspectives of Military Medical Leaders

Martin C. M. Bricknell and Stefan Kowitz

Abstract This paper provides an overview of the ethical issues to be considered by military medical leaders concerning the three inter-related topics of triage, medical rules of eligibility, and entitlement to care. The interpretation of International Humanitarian Law (IHL) and the application of medical ethics during mission medical planning is explained, alongside the processes of 'definition of population at risk', 'casualty estimation' and 'medical resource allocation'. Based on a hypothetical scenario, the paper then considers the legal, ethical, and clinical principles that might determine 'Medical Rules of Eligibility'. The paper proposes some answers to the scenario questions and concludes by discussing the importance of training and education in military medical ethics to ensure that military medical leaders do not inadvertently make unethical decisions.

Keywords Medical ethics · Law · Public health · Armed forces · Military medicine

2.1 Introduction

An effective and reliable military medical support system helps to improve troop morale and maintain the trust of military personnel, the wider public, and its political leadership. If soldiers are wounded, injured, or ill, they have to be sure that everything will be done to save their life and provide the best possible quality of recovery to either return to military duty or transition to civilian life. Every war

M. C. M. Bricknell (✉)
Conflict, Health and Military Medicine, King's College London, London, UK

S. Kowitz
Multinational Medical Coordination Centre/European Medical Command (MMCC/EMC), Koblenz, Germany
e-mail: stefankowitz@bundeswehr.org

S. M. Eagan, D. Messelken (eds.), *Resource Scarcity in Austere Environments*, Military and Humanitarian Health Ethics,
https://doi.org/10.1007/978-3-031-29059-6_2

leads to advances in military medical practice with a commensurate improvement in the probability of survival for military casualties, and translation into civilian practice (Givens et al. 2017). The recent conflicts in Iraq and Afghanistan have led to significant advances in military medical practice (Howard et al. 2019). However, these advances have led to ethical debates about access to this level of care for non-military casualties, especially local civilians. Many authors argue that ethical practice in military medicine is a key professional responsibility of all healthcare workers within military health services (Annas 2008; Wilkins and Dieppe 2017; Withnall and Brockie 2019). Military medical ethics combines the disciplines of the medical (health) professions and the military profession. This overlap is widely termed 'dual loyalty' (Olsthoorn 2019). Military medical practice covers both the deployed operational environment and garrison healthcare. In garrison, it is widely accepted that access to military medical facilities is controlled according to defined entitlements with excluded patients using other public or private facilities (Bricknell and Cain 2020).

On military operations there is a clear tension between the clinical care available to international military personnel through the whole deployed military health system (including care on return to their parent country) and that which might be available to local nationals (security forces personnel and civilians) through government or NGO provision. The control of access to military medical facilities can cause tension if the 'military necessity' of having sufficient medical resources to support military operations clashes with the 'medical humanity' to care for patients purely on the basis of clinical need (Gross 2017, 2021). This tension is summarised from a NATO perspective in the paper by Cordell (Cordell 2012). The policy for this control has been termed 'Medical Rules of Eligibility (MROE)' (Bricknell 2007) and has generated considerable debate in the academic literature (Hooft 2019; Miller 2017). The application of MROE is a key source of ethical challenge, specifically for patient-facing clinicians in military units who have to balance clinical care with the availability of empty beds for military casualties (Bernthal et al. 2017; Draper and Jenkins 2017).

The aim of this paper is to examine the issues in managing access to operational military medical facilities from the perspective of a senior medical leader as they apply to casualties from armed conflict. The term 'medical rules of eligibility' has emerged from the description of the decision pathway used as a policy framework to guide the assignment of emergency medical evacuation missions to military ambulances and helicopters by Patient Evacuation Co-ordination Centres (PECCs) (Bricknell 2007). The term may be also used to determine entitlement to access international military field hospitals for non-military patients for non-emergency care. The discussion on the topic of MROE may also include the subject of 'Triage' in the context of difficult choices in the allocation of medical resources (time, drugs, equipment, personnel). The paper will consider the relationship between all three areas, focussing on the responsibilities of the medical leader for the medical planning process that leads to the policies, orders and procedures that guide clinical decision-making. Control of access to the whole operational military health system has previously been termed 'gate-keeping' (Bricknell and dos Santos 2011). The opinions expressed reflect our personal experiences as medical leaders and planners in managing medical support of UK, Germany and coalition military operations over the past 20 years. These are complemented by reference to formal documents in the public domain and narrative descriptions of military practice from the academic literature, supported by a case example using a practical scenario.

2.2 Scenario – Case Example

This scenario provides a practical case example of the type of challenge facing senior medical leaders and planners when developing and implementing 'Medical Rules of Eligibility' policies and procedures for deployed military medical facilities. It emphasises the breadth of the stakeholders that operate in the health economy of countries experiencing complex emergencies. Whilst fictional, it is based on the practical experiences of the authors.

You are the senior medical adviser in a multinational peacekeeping headquarters that is planning the deployment of around 25,000 peacekeepers to a country that has been ravaged by civil war. The warring factions have agreed to comply with a provisional peace agreement that includes the deployment of a peacekeeping mission under an international coalition headquarters. The mission will be led by a civil-military headquarters based in the capital. There will be five subordinate mission headquarters, one to each sector. The peacekeepers will be responsible for monitoring the ceasefire within each sector and for ensuring the security of large 'protection of civilians (POC)' camps that have been set up by humanitarian organisations to meet the needs of large numbers of internally displaced persons. A camp for around 5000 peacekeepers is to be set up near each sector POC camp. The sector peacekeepers will conduct security patrols from their camp. It is assessed that there is a risk of attack from disenfranchised armed groups and criminals, a risk from unexploded remnants of war, and a risk from serious indigenous infectious diseases.

It is expected that a small field hospital (one surgical table, 10 beds, x-ray) will be established in each sector peacekeeper camp to provide medical care for peacekeepers. The sector peacekeeping force will be supported by four armoured ambulances, and a helicopter that can be configured for medical evacuation. There will be a medical escort team in the capital that can be deployed on a transport aircraft to retrieve patients from a regional camp. There is already a hospital in each POC camp run by an international NGO with limited medical capability (maternity care, surgery, x-ray). Each regional city has a civilian hospital with equally limited medical services. Peacekeepers who need to be medically evacuated from the sector camp hospital will be admitted to a multinational temporary hospital in the capital run by an international civilian commercial company on contract to the UN (with charges for all patients excluding military and UN civilians). Strategic Aeromedical

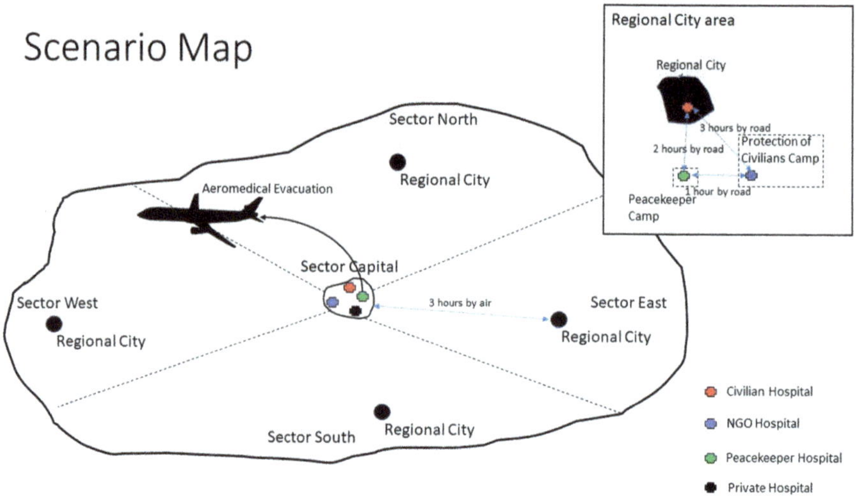

Fig. 2.1 Scenario map

evacuation from the country is the responsibility of the patient's sponsoring nation/organisation. Advanced investigations (CT scan, MRI scan, specialist laboratory tests) can be done at the leading private hospital. This is illustrated in Fig. 2.1.

You are responsible for drafting the 'Medical Rules of Eligibility' for the medical system supporting the multinational peacekeeping force that will be ratified by each of the contributing nations. Consider the following questions:

Who should be entitled to receive care at the sector field hospital?
Who should be entitled to a flight on a medical evacuation helicopter?
Who should be entitled to a flight on a medical evacuation aeroplane to the capital?
Should the sector field hospital treat any patients from the NGO hospital or the civilian hospital?

2.3 Legal, Ethical and Clinical Principles

All military medical planning is underpinned by the legal obligations of national governments under International Humanitarian Law, specifically the four Geneva Conventions and associated protocols. In most countries, these are enshrined into national law and military regulations. They invoke a duty to ensure access to medical care without distinction (except for medical priority, so called principle of impartiality) for the wounded, shipwrecked, prisoners and civilians harmed during armed conflict (Additional Protocols 1977a, b). It is the responsibility of governments to meet the healthcare needs of their populations (including governments that deploy their armed forces on military operations), and they may do this with partners including humanitarian organisations and commercial companies. In conflict or other emergencies, the United Nations may support national governments through the cluster approach to co-ordinating the international humanitarian response (Inter-agency Standing Committee 2015). There are established mechanisms for communication across the civil-military interface that are designed to protect the humanitarian principles of humanity, neutrality, impartiality and operational independence (Inter-Agency Standing Committee 2018). Within the UN system for co-ordinating the humanitarian response, the World Health Organisation has responsibility for convening the Health Cluster (Health Cluster Guide 2020). The military health system should be considered as a stakeholder in the health cluster but it is neither neutral nor operationally independent because it acts on behalf of a security actor (ICRC 2020). Reinforcing the importance of ethical practice, the International Committee of Military Medicine (ICMM) was a co-signatory to the document "Ethical Principles of Healthcare during Armed Conflict and Other Emergencies" behalf of most Surgeon Generals of the armed forces across the world (ICRC et al. 2015). As a further example of international consensus, the highest medical doctrine in NATO, Allied Joint Medical Doctrine, makes multiple references to the ethical aspects of military medical practice, including the application of MROE (NATO 2019).

In addition to their legal rights and responsibilities under IHL, military medical personnel must comply with their ethical duties as defined by their national medical bodies. These duties may be summarised by the 4 principles of medical ethics, autonomy, justice, beneficence, and non-maleficence (Beauchamp and James 2019). For this paper, the principle of justice is most important, as it covers the humanitarian principle of impartiality but also introduces the concept of fairness in the distribution of scarce resources. At the clinical level, the care provided to patients must be clinically safe (beneficence), culturally appropriate (non-maleficence), and must not cause harm to the wider health system. At the local level, a military medical facility might be clinically capable and have sufficient capacity to provide healthcare to non-military patients. However, it might be inappropriate for this facility to provide care to local nationals compared to the local clinical practice (e.g. the drugs may be different); harmful to the cultural context of the patient (e.g. men may not care for women); disruptive to the financial flows to the local health economy (e.g. the provision of free drugs may affect the income of local pharmacists), and 'contaminate' the humanitarian space for civilian health providers (e.g. medical facilities may become targets for armed groups). These risks are reflected in the guidance on the use of military resources in disaster response from the United Nations Office for the Coordination of Humanitarian Affairs (UNOCHA) Inter-Agency Standing Committee and other documents that cover the use of military medical units in disasters and other emergencies (as previously cited). Additionally, the military health care system is optimized for the needs of the military mission (trauma patients, short theatre patient return policy with rapid strategic aeromedical evacuation capacity). Clinical specialists, for example paediatrics or obstetrics, are not deployed and medical training of military health practitioners may not cover the clinical skills required to treat non-military patients. Therefore, any MROE policy that authorises access to deployed military medical facilities by local patients needs to balance a range of factors covering capability, capacity, clinical suitability, cultural appropriateness, and risks to the wider local health economy.

2.4 Policies for Entitlement to Care

Whilst many nations commit to an aspiration of Universal Health Care (UHC) based on the principles of equity and fairness, the reality is that access to healthcare is constrained by wealth, employment, social class, disability and other discriminators (Ghebreyesus 2017). Military medical services are never able to meet the full breadth of health needs of the populations in which they operate, either in garrison or on military operations. Therefore, it is inevitable that there will be discrimination for 'entitlement to care' between specific population groups based on a list of beneficiaries as defined by the 'population at risk'. This list is likely to define the range of clinical services available (possibly including medical repatriation) and any methods for payment or cost recovery. This is particularly important when nations are collaborating in 'burden-sharing' or contracting third parties to provide their medical support. The rules for 'entitlement to care' are likely to be sensitive and covered by a Memoranda of Understanding or other legal provisions between parties.

Nations will balance a range of competing factors in the allocation of resources to a military mission, including military medical services. This will be informed by the medical planning process (covered below), but the final decision will reflect wider political and economic factors. Actual provision of medical care will be constrained by 'means and capabilities' without unlimited liability. Ideally international military medical services should only provide health care for international military personnel and operate medical facilities separately from local military or civilian health actors. International efforts should focus on capacity-building of local medical services through development assistance, appropriate donation of medical material, and education; rather than direct provision of medical care. However, none of this excludes the use of military medical services in support of civilian populations affected by conflict or other emergencies as a last resort when the need cannot be met by civilian providers.

2.5 Military Medical Planning

Medical planning for specific military operations follows the wider military 'operational planning process (OPP)' that is a codified approach to solving military problems, the principles of which are used by most armed forces (Bricknell and Beardmore 2011). The process is initiated by 'mission analysis' in which the leadership team reviews the mission that they have been given to identify the subordinate tasks, freedoms, and constraints that need to be planned or considered during the OPP. This will determine the type of military operation that requires medical support and the nature of medical task. The following four pillars form the basis for medical support planning: medical service doctrine and policy, troop strength, health hazards and risks, and forecasts of casualty rates. The latter data are important for determining the necessary capacities, such as the number of operating tables, intensive care teams, beds, and ambulances.

There is a significant difference in medical planning between supporting a small-scale special operations mission and a military contribution to a large international humanitarian relief operation. In all cases, the medical planner must consider the 'population at risk'. The PAR will include national military personnel, and may include multi-national military personnel, civilians (including contractors), local security partners, prisoners or detainees, and possibly local civilians (especially if there is a humanitarian component to the mission). This determines the potential number of explicit beneficiaries of the military medical system and allows the identification of any 'implicit' beneficiaries. The next phase of medical planning is to undertake a 'casualty estimate' based on the intended plan for military operations. Based on above factors and information, software systems can estimate casualty and DNBI (Diseases, Non-Battle Injuries) rates in the sense of a benchmark demand, using experience from comparable scenarios or events of the past. The casualty estimate determines the allocation of medical resources to meet this demand from the military mission (Bricknell et al. 2011). The military medical leader is responsible for the legal, ethical and clinical oversight of the medical planning process and should seek advice, where necessary. The plan is then implemented through a series of policies, orders and standing procedures, which describe the subordinate missions, tasks, and responsibilities of all units in the military health system. This will include procedures for the application of MROE. Medical leaders are also responsible for designing the information system that provides situational awareness of the overall medical system, showing both the occupancy and future capacity, to manage the flow of patients through the system. It is likely that a multinational medical support concept for a security mission (which will be stated in orders and plans) will identify humanitarian assistance to the local civilian population as a subsidiary task within the medical capabilities and capacities assigned to support the international force (though, there may be restrictions on the use of medical equipment used for this role). The military use the abbreviation JIIM (Joint, Interagency, Intergovernmental and Multinational) to cover the breadth of stakeholders to a military plan. It will be assumed that the non-military members of the Health Cluster will take primary responsibility for addressing the health needs of the local population. Thus, the military medical services may not have any resources dedicated to meet the health needs of a civilian population (unless they are being deployed exclusively for a disaster relief mission. An example of the whole process is described in a paper on medical planning for OP MOSHTARAK, a NATO operation in Southern Afghanistan (Bricknell 2011). This responsibility for co-ordination is also outlined in the ICRC document *Protecting healthcare: Guidance for the Armed Forces* published in 2020 (ICRC 2020).

2.6 Medical Rules of Eligibility

Medical Rules of Eligibility (MROE) extend the principles for entitlement to care into a decision pathway for access to the military medical system. MROE provide decision-making rules that can be applied effectively and efficiently at the pace needed to respond to an emergency, with provision to refer difficult choices to senior leaders. They are designed to operate across the whole pathway of care, with an expectation of emergency assessment and stabilisation pending transfer from the military medical system (when clinically appropriate) for all patients who are not entitled to military health care. These are usually written to default to allowing impartial access to emergency care in accordance with IHL. However, if there is either insufficient capacity to accept the patient or a military necessity (on the authority of military commanders based on legal and medical advice) to maintain uncommitted capacity in order to support future military operations, patients might be diverted to other local medical facilities so long as this complies with IHL. MROE balances the principles of impartiality and justice with the practical aspects of allocative efficiency, recognising that the international military health system cannot meet the needs of the entire local civilian population. MROE are likely to be written for specific missions and applied to different stages in the care pathway from point of injury to reception at national referral centres. Whilst most academic discussion has covered MROE for the assignment of medical evacuation helicopters, MROE may apply to the control of entry to military compounds containing medical facilities (especially hospitals), the transfer of patients between medical facilities, and different medical conditions (as applied during the international response to Ebola in West Africa or more recently COVID-19 (Bricknell et al. 2016). MROE can only provide generic guidance and should have sufficient flexibility that the medical leadership in conjunction with the military leadership can make short-term decisions based on the clinical needs, the tactical situation, and the situation in the medical military health care system (for example occupancy of ICU beds in field hospitals and the possibility to re-distribute patients across the whole system). In spite of this, multinational military medical personnel have regularly faced ethically challenging situations during their missions due to inherent resource scarcity, hazardous conditions, conflicting cultural, doctrinal and legal factors (Bernthal et al. 2017; Draper and Jenkins 2017). Many clinical decisions about resources and prioritisation for treatment may still need to be taken at short notice.

2.7 Triage

The preceding discussion on MROE covers policy for the allocation of medical resources, without constraining the professional duties of patient-facing healthcare professionals. Triage covers the process of deciding the order for treating individual patients in an emergency, based on clinical need underpinned by the principles of impartiality and humanity without distinction. Triage is governed by the clinical assessment of patients as described by procedures for the management of a Major Incident with large number of casualties (Falzone et al. 2017). In normal circumstances, casualties are placed into one of 3 triage categories (T1 – immediate, T2 – urgent, T3 – minimal). Military medical services also consider the conditions of a mass casualty situation (MASCAL) in which the number, type or severity of casualties exceeds the treatment capacities and capabilities available. In these circumstances, the priorities for triage may be changed from prioritisation based on severity of injury to prioritisation based on achieving the best outcome for the greatest number of patients. This introduces an additional triage category (T4 – expectant) (NATO 2019). This shifts the allocation principle to utilitarianism, 'the greatest good for the greatest number'. The decision to declare a MASCAL is held at a very senior level because it implies that the available medical resources have been exceeded. When the MASCAL situation is under control, the T4 category will stop being used. The recent COVID-19 pandemic brought this aspect of triage into the civilian domain with the development of ethical guidelines to guide clinicians facing difficult decisions about the allocation of intensive care beds (Ehni et al. 2021; Tyrrell et al. 2021; Vinay et al. 2021). Following this experience, it may be appropriate to introduce a formal 'ethics oversight board' to review the MROE for specific military operations.

2.8 Review of the Scenario Questions

Returning to the scenario, there is likely to be a complicated network of stakeholders involved in meeting the health needs of all populations affected by the complex emergency. The medical facilities deployed to support the multinational peacekeeping force are very limited in both capacity and capability. They are specifically resourced and configured to support international military forces. The local national and displaced populations already have access to an established network of government, NGO, and commercial health services though there may be a significant gap between these capabilities and their health needs. Based on the scenario, including the availability of non-military health services, an indicative MROE matrix is shown at Table 2.1. The left column shows the levels of care and the top row shows categories of patient that might be included in the Population at Risk (PAR). Military and civilian members (if agreed in the operational planning documents) of the peacekeeping force are likely to be entitled to all levels of care in accordance with the medical plan. There may be specific arrangements for covering the costs of commercial provision including health care at the private hospital between these groups. This meets the duties of the sending states to provide health services to the sick and injured of their armed forces under the First Geneva Convention (1949) and the ethical obligation of governments to provide support to their armed forces. Local civilians employed in the camp are likely to be entitled to emergency treatment whilst they are in the camp (e.g. from an accident) but not routine medical care, this meets

Table 2.1 Indicative medical rules of eligibility for access to deployed military healthcare

Levels of care/Category of PAR	Military peacekeepers	Civilian peacekeepers	Local civilians in the sector camp	Local civilians harmed in an incident involving peacekeepers	Local civilians from outside the camp	Locals detained by peacekeepers
Care at the sector field hospital	✓ unrestricted	✓ unrestricted	✓ emergency only, and transfer	✓ emergency only, and transfer	✗	✓ unrestricted
Flight on a sector medical evacuation helicopter	✓ unrestricted	✓ unrestricted	Ground ambulance only	✓ emergency only, and transfer	✗	✓ unrestricted
Flight on a medical evacuation aeroplane to the capital	✓ unrestricted	✓ unrestricted	✗	✗	✗	? if they require a higher level of care
Admission to the civilian private hospital in the capital	✓ unrestricted	? if this has already been agreed	✗	✗	✗	✗

the principles of the Additional Protocols (Additional Protocols 1977a, b) and the ethical duty to provide emergency healthcare without distinction. They would be transferred to the local civilian system and are unlikely to be entitled to medical evacuation by aircraft to the next level military facility. Similarly, local civilians hurt in an incident involving peacekeepers may be entitled to a flight on an emergency helicopter under the obligation to provide access to medical care, though they might be taken to a civilian hospital. It is unlikely that military emergency helicopters would be sent to an incident not involving peacekeepers as the lack of security at the landing zone might incur too great a risk to the aircraft. Local civilians would not be entitled to emergency or other forms of care from the camp hospital as alternative civilian medical facilities are available, though there might be arrangements for a clinical review of emergency cases at the camp entrance to check that they will survive transport to the other civilian medical facilities. Finally, there will be specific arrangements for the provision of healthcare to any person detained by the peacekeeping force, based on political guidance in cooperation with host nation. No detainee will be transferred from a military hospital, if it will cause risk to the health of the detainee. This complies with the Third Geneva Convention (1949).

This table would be used as the basis for policy for MROE within the international peacekeeping force, based on the questions from the scenario. However, individual circumstances will remain that require clinical assessment and judgement. These might include: the evaluation of a severely injured patient at the front gate who will not survive transport to another facility; the transfer of a seriously ill civilian patient from an incident involving peacekeepers after initial treatment if the local medical facilities are not capable of providing the same level of care; the use of blood in the camp field hospital to treat local nationals; or the management of care for an international contractor if their employer does not provide access to the private hospital in the capital. The COVID-19 crisis has brought additional factors into the analysis of entitlement to care across peacekeeping and humanitarian missions, particularly pertaining to access to tests, vaccinations and aeromedical evacuation (Zhang et al. 2020).

The final question in the scenario considered the circumstances of a referral from the NGO or the local civilian hospital for treatment by the camp field hospital. This might occur if one of the international clinicians had expert skills beyond the capability of the local doctors (e.g. maxillofacial, plastic, or orthopaedic surgery). This might be an appropriate referral if arrangements had been previously agreed, including full assessment of the patient. In order to avoid using a military hospital bed, the military surgeon might perform the procedure in one of the civilian facilities if the security situation and wider clinical capability of the hospital were suitable. This possibility will only work, if adequate and sufficient follow-on treatment including rehabilitation is guaranteed and planned.

2.9 Training of Medical Personnel

The experience of the last two decades, reflected in the range of supporting references to this paper, emphasise importance of training and education in IHL and medical ethics for military health personnel. This should be based on establishing a baseline foundation of knowledge through initial professional training and reinforced during clinical and military training, especially field exercises. Training should involve small group discussions based around practical scenarios from real events such as those available from the King's Centre for Military Ethics (Miron and Bricknell 2021) or the ICMM Reference Centre in Military Medical Ethics (Messelken 2018). Military commanders should also be taught the key elements of military medical ethics as part of their training in the laws of armed conflict to prevent them issuing an unlawful order to military medical personnel. Senior military medical leaders should be actively involved in the interpretation of IHL and medical ethics for military health policy, especially with key external interlocutors such as the ICRC and human rights lawyers.

2.10 Conclusions

The medical leader is responsible for oversight of the military health system covering: the planning process; issuing policy, orders and procedures; and managing patient flow through the system. They must ensure that this is conducted in a legal and ethical manner, alongside providing safe and effective clinical care to patients treated within the military health system. They should specifically consider the legal and ethical implications of policies and procedures for MROE. This requires a balance between eligibility for treatment in military medical facilities that are optimised to rapidly treat and evacuate international military patients with the ethical and legal duty to enable the *'wounded, sick and shipwrecked to receive, to the fullest extent practicable and with the least possible delay, the medical care and attention required by their condition. There shall be no distinction among them founded on any grounds other than medical ones'* (Additional Protocols 1977a, b). However, these conventions require the medical leader to be sufficiently trained, experienced, and equipped with methods of leverage (in respect of the military leadership) to find this balance and implement necessary means to reinforce it. The MROE should be designed to enable delegated decision-making, and include access to senior advice for difficult cases. Given the sensitivity of these policies, MROE might require review by external experts. In doing this, military medical leaders are balancing the obligations of both the healthcare and the military professions. Necessarily, this entails that they will discuss the MROE with their non-medical military peers. The scenario in this paper has provided an insight to some of nuances of such policies and procedures.

References

Annas, G.J. 2008. Military medical ethics — Physician first, last, always. *New England Journal of Medicine* 359 (11): 1087–1090. https://doi.org/10.1056/nejmp0805975.

Beauchamp, T.L.C., and F. James. 2019. *Principles of biomedical ethics*. 8th ed. New York: Oxford University Press.

Bernthal, E.M., H.J.A. Draper, J. Henning, and J.C. Kelly. 2017. 'A band of brothers'—An exploration of the range of medical ethical issues faced by British senior military clinicians on deployment to Afghanistan: A qualitative study. *Journal of the Royal Army Medical Corps* 163 (3): 199–205. https://doi.org/10.1136/jramc-2016-000701.

Bricknell, M. 2007. Reflections on medical aspects of ISAF IX in Afghanistan. *Journal of the Royal Army Medical Corps* 153 (1): 44–51. https://doi.org/10.1136/jramc-153-01-11.

———. 2011. Medical lessons from OPERATION MOSHTARAK phase 2. *Journal of the Royal Army Medical Corps* 157 (Suppl 4): S463–S467. https://doi.org/10.1136/jramc-157-4s-12.

Bricknell, M., and C. Beardmore. 2011. The context for planning military health services support. *Journal of the Royal Army Medical Corps* 157 (Suppl 4): S430–S433. https://doi.org/10.1136/jramc-157-4s-04.

Bricknell, M., and P. Cain. 2020. Understanding the whole of military health systems. *The RUSI Journal* 165 (3): 40–49. https://doi.org/10.1080/03071847.2020.1784039.

Bricknell, M.C., and N. dos Santos. 2011. Executing military medical operations. *Journal of the Royal Army Medical Corps* 157 (Suppl 4): S457–S459. https://doi.org/10.1136/jramc-157-4s-10.

Bricknell, M.C., F. Jones, and J. Hatzfeld. 2011. Casualty estimation and medical resource planning. *Journal of the Royal Army Medical Corps* 157 (Suppl 4): S439–S443. https://doi.org/10.1136/jramc-157-4s-06.

Bricknell, M., T. Hodgetts, K. Beaton, and A. McCourt. 2016. Operation GRITROCK: The Defence medical Services' story and emerging lessons from supporting the UK response to the Ebola crisis. *Journal of the Royal Army Medical Corps* 162 (3): 169–175. https://doi.org/10.1136/jramc-2015-000512.

Civil-military coordination during humanitarian health action (provisional). 2011. Retrieved from https://www.who.int/publications/i/item/civil-military-coordination-during-humanitarian-health-action

Convention (I) for the Amelioration of the Condition of the Wounded and Sick in Armed Forces in the Field., 12 August 1949.

Convention (III) relative to the Treatment of Prisoners of War., 12 August 1949.

Cordell, R. 2012. Multinational medical support to operations: Challenges, benefits and recommendations for the future. *Journal of the Royal Army Medical Corps* 158 (1): 22–28. https://doi.org/10.1136/jramc-158-01-06.

Draper, H., and S. Jenkins. 2017. Ethical challenges experienced by UK military medical personnel deployed to Sierra Leone (operation GRITROCK) during the 2014–2015 Ebola outbreak: A qualitative study. *BMC Medical Ethics* 18 (1): 77. https://doi.org/10.1186/s12910-017-0234-5.

Ehni, H.-J., U. Wiesing, and R. Ranisch. 2021. Saving the most lives—A comparison of European triage guidelines in the context of the COVID-19 pandemic. *Bioethics* 35 (2): 125–134. https://doi.org/10.1111/bioe.12836.

Falzone, E., P. Pasquier, C. Hoffmann, O. Barbier, M. Boutonnet, A. Salvadori, et al. 2017. Triage in military settings. *Anaesthesia Critical Care & Pain Medicine* 36 (1): 43–51. https://doi.org/10.1016/j.accpm.2016.05.004.

Ghebreyesus, T.A. 2017. All roads lead to universal health coverage. *The Lancet Global Health* 5 (9): e839–e840.

Givens, M., A.E. Muck, and C. Goolsby. 2017. Battlefield to bedside: Translating wartime innovations to civilian emergency medicine. *The American Journal of Emergency Medicine* 35 (11): 1746–1749.

Gross, M.L. 2017. Saving life, limb, and eyesight: Assessing the medical rules of eligibility during armed conflict. *The American Journal of Bioethics* 17 (10): 40–52. https://doi.org/10.1080/15265161.2017.1365186.

———. 2021. *Military medical ethics in contemporary armed conflict: Mobilizing medicine in the pursuit of just war*. Oxford University Press.

Health Cluster Guide. A practical handbook. 2nd Edition. 2020. World Health Organisation, Geneva. Available at: https://healthcluster.who.int/publications/i/item/9789240004726

Hooft, F.B. 2019. Legal framework versus moral framework: Military physicians and nurses coping with practical and ethical dilemmas. *Journal of the Royal Army Medical Corps* 165 (4): 279–281. https://doi.org/10.1136/jramc-2018-001137.

Howard, J.T., R.S. Kotwal, C.A. Stern, J.C. Janak, E.L. Mazuchowski, F.K. Butler, et al. 2019. Use of combat casualty care data to assess the US military trauma system during the Afghanistan and Iraq conflicts, 2001–2017. *JAMA Surgery* 154 (7): 600–608. https://doi.org/10.1001/jamasurg.2019.0151.

ICRC. 2020. *Protecting healthcare: Guidance for the armed forces*. Geneva: ICRC. Retrieved from https://www.icrc.org/en/publication/4504-protecting-healthcare-guidance-armed-forces.

ICRC, WMA, ICMM, IFMSA, FIP, WCPT. 2015. *Ethical principles of health care in times of armed conflict and other emergencies*. Retrieved from https://www.icrc.org/en/document/common-ethical-principles-health-care-conflict-and-other-emergencies

Inter-Agency Standing Committee, I. T. I.-A. S. 2015. Reference module for cluster coordination at the country level. In: Geneva: IASC (https://www. humanitarianresponse. info/sites/www ….
———. 2018. *Recommended practices for effective civil-military coordination of foreign military assets (FMA) in natural and man-made disasters* (version 1.0). Retrieved from https://www.unocha.org/sites/unocha/files/180918%20Recommended%20Practices%20in%20Humanitarian%20Civil-Military%20Coordination%20v1.0.pdf
Medical Aspects in the Management of a Major Incident/Mass Casualty Situation. 2015. NATO Retrieved from https://www.coemed.org/files/stanags/03_AMEDP/AMedP-1.10_EDA_V1_E_2879.pdf
Messelken, D. 2018. A collection of scenarios in military medical ethics to support training and research. *International Review of the Armed Forces Medical Services* 91 (4): 75–77.
Miller, J.P. 2017. A care ethics approach to medical eligibility in armed conflict. *The American Journal of Bioethics* 17 (10): 61–63. https://doi.org/10.1080/15265161.2017.1367867.
Miron, M., and M. Bricknell. 2021. Innovation in education: The military medical ethics 'playing cards' and smartphone application. *BMJ Military Health*. bmjmilitary-2021-001959
NATO. 2019. *Allied joint doctrine for medical support*. NATO standardisation office (NSO) Retrieved from https://www.coemed.org/files/stanags/01_AJP/AJP-4.10_EDC_V1_E_2228.pdf
Olsthoorn, P. 2019. Dual loyalty in military medical ethics: A moral dilemma or a test of integrity? *Journal of the Royal Army Medical Corps* 165 (4): 282–283. https://doi.org/10.1136/jramc-2018-001131.
Protocol Additional to the Geneva Conventions of 12 August 1949, and Relating to the Protection of Victims of International Armed Conflicts (Protocol I), June 8, 1977a. Retrieved from https://ihl-databases.icrc.org/applic/ihl/ihl.nsf/Treaty.xsp?documentId=D9E6B6264D7723C3C12563CD002D6CE4&action=openDocument
——— to the Geneva Conventions of 12 August 1949, and relating to the Protection of Victims of Non-International Armed Conflicts (Protocol II), 8 June 1977b. Retrieved from https://ihl-databases.icrc.org/applic/ihl/ihl.nsf/7c4d08d9b287a42141256739003e636b/d67c3971bcff1c10c125641e0052b545
Tyrrell, C.S.B., O.T. Mytton, S.V. Gentry, M. Thomas-Meyer, J.L.Y. Allen, A.A. Narula, et al. 2021. Managing intensive care admissions when there are not enough beds during the COVID-19 pandemic: A systematic review. *Thorax* 76 (3): 302–312. https://doi.org/10.1136/thoraxjnl-2020-215518.
Vinay, R., H. Baumann, and N. Biller-Andorno. 2021. Ethics of ICU triage during COVID-19. *British Medical Bulletin* 138 (1): 5–15. https://doi.org/10.1093/bmb/ldab009.
Wilkins, D., and C. Dieppe. 2017. The non-combatant status: Importance and implications for medical personnel. *Journal of the Royal Army Medical Corps* 163 (6): 366–370. https://doi.org/10.1136/jramc-2016-000732.
Withnall, R., and A. Brockie. 2019. Military ethics: An operational priority. *Journal of the Royal Army Medical Corps* 165 (4): 219–219. https://doi.org/10.1136/jramc-2019-001262.
Zhang, Y., D. Xiang, and N. Alejok. 2020. Coping with COVID-19 in United Nations peacekeeping field hospitals: Increased workload and mental stress for military healthcare providers. *BMJ Military Health.*, bmjmilitary-2020-001642. https://doi.org/10.1136/bmjmilitary-2020-001642.

Professor Martin C. M. Bricknell was appointed as Professor of Conflict, Health and Military Medicine at King's College London in April 2019. Prior to this he served 34 years in the UK Defence Medical Services, culminating his service as the Surgeon General of the UK Armed Forces. He undertook operational tours in Afghanistan, Iraq, and the Balkans with multiple additional overseas assignments. In 2010 and 2006, he held senior Medical Adviser appointments in the NATO ISAF mission. He was awarded the Companion of the Order of Bath, the Order of St John and the US Bronze Star during his military service. He is an accredited specialist in General

Practice, Public Health and Occupational Medicine. His multiple academic papers cover: how organisations learn, care pathways in military healthcare, military healthcare ethics, civil-military relations in health, and the political economy of health in conflict. He is also Deputy Director of the KCL Centre for Military Ethics, Veterans Adviser for the King Edward VII hospital, Editor-in-Chief of the Military Medical Corps Worldwide Almanac, a non-resident Fellow of the Centre for Global Development, and on the editorial boards for the Journal of Military and Veterans Health and BMJ Military Health.

Brigadier General (MC) Dr. med. Stefan Kowitz took up his appointment as Director, MMCC/EMC in Koblenz/Germany in August 2019 to build up, together with his team, this new entity, supporting NATO and EU in medical affairs. Prior to this, he held the highest medical position in NATO: Medial Advisor for the Supreme Headquarters Allied Powers Europe (SHAPE), Mons/BEL. From 2014 to 2016, he was able to collect experience as Director of the NATO Centre of Excellence for Military Medicine, Budapest/HUN. During that time, he was member of the Scientific Advisory Board of the Department of Military, Disaster and Law Enforcement Medicine (Faculty of Medicine of Semmelweis University Budapest). He was deployed in 1995 as Deputy Commander of German Field Hospital, TROGIR, HRV; in 2005 as Medical Director and Advisor at HQ ISAF KABUL, AFG and in 2011 as Commander of the German Medical Task Force Afghanistan. He was awarded the Federal Cross of Merit on Ribbon, Hungarian Service Medal of Merit in Gold. He is a general practitioner and an accredited specialist in both Occupational and Sports Medicine. He is especially interested in management in military healthcare, and the political economy of health in crisis and conflict.

Chapter 3
Reconsidering Triage: Medical, Ethical and Historical Perspectives on Planning for Mass Casualty Events in Military and Civilian Settings

Simon Horne, Robert James, Heather Draper, and Emily Mayhew

Abstract A mass casualty (MASCAL) event is different to a major incident. The crux of this difference is that in a major incident, by the adoption of special measures, normal or near-normal standards of care can be maintained. In a MASCAL, irrespective of what special measures are instituted, standards of care inevitably drop. This is a, currently unmet, challenge for medical planning and planning policy. Twenty-First century weaponry is capable of producing thousands of causalities a day over a period of several days in peer-on-peer conflict. In this chapter, we propose that medical planning for military events on this scale should include the following: explicit acceptance that 'gold standard' care cannot be given to patients who are triaged for treatment, a better understanding how to identify those patients who will not be triaged for treatment, a focus on ensuring patient flow (which includes diversion, self-care and remote monitoring), and implementation of the principle that care should be given by the lowest capable provider. We attempt to begin to develop some principles for planners to consider drawing on historical precedents and explore some of the ethical implications of our proposals.

S. Horne
Centre for Defence Healthcare Engagement, Royal Centre for Defence Medicine, Birmingham, UK
e-mail: simon.horne@nhs.net

R. James
Academic Department of Military Emergency Medicine, Birmingham, UK
e-mail: robertjames1@nhs.net

H. Draper (✉)
Warwick Medical School, University of Warwick, Coventry, UK
e-mail: h.draper@warwick.ac.uk

E. Mayhew
Imperial College, London, UK
e-mail: e.mayhew@imperial.ac.uk

S. M. Eagan, D. Messelken (eds.), *Resource Scarcity in Austere Environments*, Military and Humanitarian Health Ethics,
https://doi.org/10.1007/978-3-031-29059-6_3

Keywords Triage · Military medical ethics · Mass casualty events

3.1 Introduction

Planning for medical support to military operations has to encompass a number of different multiple casualty scenarios. Traditional doctrine recognises that such events can range from Major Incidents (MIs) to Mass Casualty (MASCAL) incidents. The crucial differentiation is that in major incidents, with especial effort, the system has the capacity to provide normal or near-normal standards of care. In MASCAL incidents, the number or nature of casualties or concomitant disruption to response infrastructure leads to the health system being overwhelmed, despite all efforts and a drop in the standard of care is inevitable.

Current NATO doctrine (North Atlantic Treaty Organisation (NATO) 2015) considers MIs and MASCAL incidents as parts of a spectrum and requires that they be managed in largely the same way. In the UK this management is typically based on the Major Incident Medical Management and Support (MIMMS) Course (Advanced Life Support Group 2011) and the Joint Emergency Services Interoperability Principles (JESIP) programme (Joint emergency services interoperability principles (JESIP) 2016), which was introduced following failings identified from response to previous civilian major incidents. In this paper we argue that while these systems are very effective in the management of MIs, they have no solutions for the unique problems of a MASCAL.

In an attempt to develop possible solutions for the extremely challenging problem of MASCAL management, we first provide evidence that standards of care can not be maintained and that a different approach is required. We then examine contemporary and historical examples to derive inspiration for how we could begin to manage a MASCAL situation. We look at the current overmatch faced by the National Health Service (NHS) in the UK; which we believe meets our definition of a MASCAL, albeit a chronic rather that acute one. We investigate some of the strategies employed by NHS urgent and emergency care networks and explore their utility in the military setting. We also review the innovations delivered on the Western Front during the First World War, in an attempt to manage MASCALs, and again try and use those principles to help inform modern military medical planning. Finally, but importantly, we review the ethical arguments relating to our proposed changes to MASCAL management. While many of these lessons may be transferable to a range of contexts, for the purposes of this paper we are examining solely military, mass casualties arising from peer-on-peer conflict.

3.1.1 *Why MASCAL Management Needs to Be Rethought*

Despite the definition of a MASCAL explicitly stating that the system will be overwhelmed, the Major Incident models are all that practitioners are given to work with. No additional strategies are given to guide management once the system succumbs to the overwhelming weight of casualties. Those MI models all work on a tacit assumption that if resources are carefully managed and directed to those most in need, there will eventually be enough to do what needs to be done i.e., the incidents in question are always compensated. Recent major incidents in Western societies have reinforced this assumption – with the large numbers of casualties generated by terrible events such as the Madrid train bombings (193 killed, approximately 2000 injured) (Wikipedia contributors 2004) or Las Vegas shooting 2017 (60 killed, 867 injured) (Wikipedia contributors 2017) being managed across region-wide hospital networks and near normal standards of care supposedly being maintained. Indeed, the coroner leading the inquest into the 7/7 bombings in London (56 killed, 784 injured) explicitly stated that a MI was no excuse for a reduction in clinical standards (Home Office 2005).

MASCAL incidents will, by definition, fall outside that scope – crossing a threshold beyond which existing systems cannot cope. Incidents demonstrating this point have occurred, but while referenced, have not obviously influenced Western mass casualty doctrine. Examples of incidents where maintenance of ususal clinical standards is not a realistic aspiration include the Bhopal gas leak (1984) in which it was estimated 3900 people suffered immediate, severely disabling injuries (Wikipedia contributors 2021a), or the San Juanico liquid propane explosion with around 6000 suffering severe burns (Wikipedia contributors 2021b). More recently, the fertiliser explosion in Beirut (2020) was estimated to have injured 7000 (BBC news 2020). An account from an Emergency Medicine Registrar that day describes how:

> Triage was set up outside the ED. People were everywhere, screaming, searching for loved ones. Some tried to get in using physical force. Many succeeded. We got hit. We got hurt. We carried on. Registration was not working, and our victims had no names… We ran out of medical supplies and the same blade was used, gloves were not changed, staples were re-used. We just wanted our patients to live. (Abdul Nabi and Sawaya 2020)

An editorial relating to that incident also describes how cardiac arrests were managed differently before and during the disaster, and the ethical issues arising from that 'change in care'. This changing standard of care, mandated by the situation, describes perfectly what happens when a system has decompensated. At some point, no matter how good the triage or the surgeon, standards of care cannot be maintained. Within current constructs, when that point is reached standards will drop precipitously, (Kipnis 2003) as there is no 'Plan B'. Individuals will be forced to adapt the care that they deliver in real time with no framework, trying to reconcile what resources are left with what demand remains and making *ad hoc* decisions about what is feasible and ethical. This drop in standards is not as simple as just changing the threshold for when a given procedure such as CPR is abandoned as futile – it reflects changes in the capacity to perform all aspects of care. There is no longer enough space to triage patients, staff to monitor for deterioration nor resources to undertake necessary surgery. From the brief extract above it can be seen that normal standards of infection control, documentation and governance all disappeared. Critically, this is not for lack of training on the part of clinicians. It is a system failing, their plans simply will not have conceived of a situation where supply was so overmatched by demand.

Military officers involved in MASCAL planning need to recognise that in contrast to operations undertaken by Western militaries in the early part of the twenty-first century, conflict-related mass-casualties beyond the capacities of battlefield health systems are again possible, and must be planned for. The Russian missile salvo against 2 Ukrainian Battalions in Zelenopillya (2014) saw 93 wounded (and all their equipment destroyed) in under 5 min (Watling 2019). Planners are now being forced to look at how casualties in the hundreds or even thousands per day could be managed, especially if this were to continue over days or weeks. In other words, military medical planning needs to extend its scope to include not only numbers that can be managed in a distributed network, still (in theory) able to provide gold-standard care to every patient, but also to encompass situations where a chronically fatigued, and potentially targeted, healthcare system is consistently overwhelmed. We have been here before – in World War One, one dressing station saw over 1000 casualties in 48 h (The History Press 2021), and in World War Two in Bari over 600 casualties were sustained after one 20-min air raid caused a gas release (Reminick 2001). Clearly Hiroshima and Nagosaki took these 'normal' war contexts to a completely different level again.

The above demonstrates that a true MASCAL incident is a realistic possibility for deployed military healthcare systems. We have also suggested that it is inevitable that clinical standards will fall, potentially catastrophically, in a MASCAL incident. We will now examine our first case study, the NHS in the UK.

3.2 Case Study 1 – The NHS

3.2.1 Evidence from the UK National Health Service That Normal Standards Cannot Be Maintained in Times of Extreme Pressure

Staff who work in UK NHS emergency departments (EDs) are used to a system that is regularly overmatched. Demand has increased year on year, so that even before the coronavirus (COVID-19) pandemic, hospitals were often at or above capacity on a daily basis. The total number of beds available to the NHS has halved from around 299,000 in 1987/88 to 141,000 in 2018/9, while the number of patients treated has increased significantly (Ewbank et al. 2017). In 2018/19, overnight bed occupancy averaged 90.2%, and regularly exceeded 95 per cent in winter. This lack of capacity has been closely associated with crowding of UK EDs and a subsequent deterioration of performance against national targets. More importantly, overcrowding is clearly associated with actual patient harm. Recent reports have demonstrated an increase in mortality rates of up to 8%, with the greatest increase occurring in the region with the worst crowding (Calver and Lintern 2022). This corroborates an older study that found a 5% increase in all-cause mortality in crowded EDs (Sun et al. 2013). In US studies, only 67% of patients with a time-critical illness were seen within recommended timelines. This has a knock-on effect of poorer clinical outcomes. Other studies have focussed on specific conditions, for example a doubling of adverse events for acute coronary syndromes from 3% to 6% in crowded departments (Pines et al. 2009). A weight of evidence again leads us to two inescapable conclusions:

1. Gold standard care cannot be achieved in an overwhelmed system.
2. Harm will result.

This is true in a Western, high-capacity healthcare system working under its normal levels of demand. In a military mass casualty incident, it is therefore inevitable that standards cannot be met – leading to a clear ethical conundrum:

> Knowing that the system will not have the resources to manage, is it better to *intentionally lower standards* in a planned way that leads to the best possible outcome, in the context of the MASCAL incident, or to *persist in the pursuit of gold standard care* until you simply run out of resources (be that time, space, personnel or equipment) and then provide care *ad hoc*?

Having again demonstrated the implications of a MASCAL incident we will now review some solutions that the NHS has tried to implement.

3.2.2 NHS Adaptations to Extreme Pressure (Analogous to a MASCAL Incident)

The NHS and similar systems have been exposed to analogues for demand:resource mismatch for a prolonged period and have been forced to adopt non-traditional solutions in order to mitigate overcrowding and resource depletion. These include:

- A move to the concept of flow to maintain operational capacity, and reduction in emphasis on 'Triage for Priority'.
- Strategies to maximise flow, such as diversion of patients, utilising alternative, lower-level providers, self-care and remote monitoring.
- Recognising that in extreme situations (such as pandemics), the pursuit of gold standard care for all is not feasible and scarce resources must be distributed for best effect.

3.2.2.1 From 'Triage for Priority' to Flow

Conventional principles of MI management follow a simple hierarchy of priorities. First, establishing command and control, ensuring rescuer safety (to prevent further casualties), then saving lives already at risk, then preventing deterioration and alleviating suffering.

Generally, casualties are triaged into order of priority (need), and then resources allocated to them on this basis sequentially. Only once those whose lives are at risk are stabilised, can operating capacity then be used to treat those with urgent but not immediate needs, and finally, after them, those who are stable but need treatment. These groups are not supposed to be ignored during this process – rather they are placed in holding areas where they could be observed for deterioration, re-triaged and be treated for pain etc. How realistic this expectation of ongoing care and observation is in the context of a completely overwhelmed system is not addressed. Moreover, in the existing paradigm, there is an acknowledgement that some patients are so severely injured that there is little or no-likelihood of recovery, and that in the event that demand significantly exceeds the resources available, these such patients may be labelled 'expectant and active treatment withheld' in order to concentrate on those more likely to survive. Again, there is an expectation that these patients will still receive some care during this period although again how feasible it is to care for those unlikely to recover whilst totally overwhelmed, is not known. Anecdotal reports suggest they receive little or no care. Whilst the formal declaration of the use of the 'futile' category is extremely rare (in the UK it requires the Secretary of State for Health's direction), (Emergency Preparedness RaR 2017) individual clinicians may make a decision in the 'patient's best interest' that aggressive management is inappropriate, just as they would in their normal practice. It should be noted that the best interests in this case may be of all the patients, not just the one in front of the clinician at that point in time.

There are three key problems with the physiological triage that underpins the current process for prioritisation of patients: batching, storage and inaccuracy. We will now address each of these in turn.

Batching is well recognised in systems research as an inefficient way to use resources, and the NHS England Urgent Care Checklist specifically identifies reduction of batching as a way to improve patient flow (NHS interim management and support 2011). While not all concerns about batch production (found in the manufacturing literature) are relevant to this discussion of MI management, some of them certainly are. The first disadvantage applicable here is the requirement to store the 'product', pending definitive treatment, whilst the rest of the batch is completed. When translated into MI triage this means that there will need to be physical space for patients to be 'stored'. In a MASCAL event, as seen in Beirut, this space will become a very real constraint. In UK emergency departments the focus has shifted to 'flow'. Patients are started on a pathway that sees them receiving a package of care started as early as possible. There is no requirement for all other patients to have been triaged before this care is commenced. Patients are then moved through the system to the most appropriate place to await further management. This could be a ward, or a clinic, or home but it is most definitely NOT remaining, untreated, in an assessment area. It is waiting untreated that is the second significant disadvantage of batching. While MI doctrine states that patients waiting will be cared for, monitored and given necessary stabilising treatments, the experience from Beirut clearly showed that this is not a safe assumption. Indeed, not only are they not receiving their definitive treatment, in a system that has, by definition, exceeded its ability to cope they may not be receiving any care at all. Given that one of the important factors when treating seriously ill and injured patients is minimising time to intervention, it is obvious that this is undesirable.

The main advantage of batching is that it allows quality control at each stage of a complex manufacturing process, but clearly that is not relevant in a mass casualty setting. Where resources are saturated there is no prospect of having staff available to audit the effectiveness of the triaging 'process' or the eventual treatment provided (i.e. conducting quality control). At best, low-level monitioring may be possible to determine whether a patient's status has changed whilst they are awaiting treatment, although, conversely a change in patient status does not mean that triage was wrong. It may simply reflect the progression of disease, that is inevitable in those not being treated. Audit in this setting would also be flawed as there is no clear standard of care for MASCAL on the scale envisaged – a problem we will be returning too. At best, sufficient data may be available to do a rough and ready outcomes comparison and a 'lessons learned for next time' exercise undertaken. Improving the quality of the process and eventually treatment during the MASCAL, given the absolute scarcity of resources, is likely to be done purely anecdotally by the staff at the front line working, on a trial and error basis.

It should also be noted that in manufacturing, batching is specifically recommended when making small numbers in each batch. This is perfect for smaller MIs such as a coach crash, but is not the situation found in a MASCAL.

A further potential advantage of batching is that it allows one person or a small group of people to focus on a specific task, it allows them to become more expert at that task and not be distracted from it. However, given the frequency of MASCAL, it is unlikely that said expertise will ever be used again by that individual.

Finally, having all patients triaged prior to any treatment decisions potentially allows the best decision to be made in terms of effective distribution of resources, albeit not necessarily in the quickest timescale. However, the reality of a MASCAL incident is that there are likely to be dozens or even hundreds of patients who need immediate treatment in order to survive. Thus, waiting for all patients to be triaged in order to increase efficiency is likely to result in the death of some patients who would have been salvageable had they been treated immediately. Saving these lives seems a reasonable trade of for a slightly less efficient solution, or risking treating some patients slightly out of order.

It should also be noted that physiological triage sieves (as recommended currently) are unable to differentiate patients with the degree of granularity required to make complex treatment decisions. They also take too long. Even at 30 s per patient, triaging 10 patients takes 5 min. Anecdotal evidence suggests that at various MIs, e.g. the London 7/7 bombings, anatomical triage was used. These anatomical methods of triage require a senior clinician to make clinical judgements based on the patient in front of them. Whilst more demanding in terms of the requirement for clinical expertise and less readily reproducible, there is evidence that anatomical triage may be superior to physiological triage in reducing mortality (Montán et al. 2011).

Because of these problems, the World Health Organisation (WHO) is considering abandoning protocolised triage systems in the initial phases of an incident, moving towards a senior clinical decision-maker making a clinical judgement on who has priority for treatment rather than following an algorithm. This reinforces the need to treat the patient in front of you as quickly as possible to make way for the next.

Recognising the implications of these three key features of current triage processes – inaccuracy, batching and storage – it becomes clear that when there is no space to store, and sick patients will die if not treated as they are found, systems must instead concentrate on *flow*. Patients must immediately be placed on a pathway that delivers only the treatments they need i.e. those that prevent them dying and possibly in the miltary context those that allow them to return to duty. This treatment must also take place in the most appropriate place (ideally NOT the Role 2 Medical Treatment Facility (MTF)) as quickly as possible. This may require them to be diverted away form the nearest receiving MTF either before or after initial treatment. Some may be diverted successfully by current processes (e.g. T/P3, also known as 'walking' or 'delayed' casualties, are usually talked about as being managed elsewhere, such as the primary care area). However, for MASCAL management to be successful we believe that we must become better at diverting patients away from the closest receiving MTF in order to prevent them inadvertently obstructing the system, see below.

3.2.2.2 Strategies to Maximise Flow: Diversion/Remote Monitoring

When considering the numbers of casualties being discussed here, it is highly unlikely they can all be admitted to one facility, particularly a small facility such as a deployed MTF. These might typically have two resus bays, one operating table, two ICU beds and twelve ward beds (typically written as 2/1/2/12). It will, therefore, be necessary to divert patients away from the primary MTF, even if it is the only one within a reasonable distance of the incident. This altered decision-making around the receiving facility has historically occurred in purely civilian MIs, but it has been within the context of a well-established health network with multiple alternative treatment facilities. For example, patients were not necessarily transferred to the closest hospital during the London 7/7 bombings. The additional travel time was, however, minimal. With a single, overwhelmed facility, it will not be possible to keep all patients at that facility, even if extraordinary measures are instituted.

One option is to provide a means of mass transport to a remote MTF. This mimics the approach taken with hospital trains during the First World War (Bricknell 2003) and seen again in the Ukranian conflict. However, this does raise a further question regarding how patients are selected for onward transfer rather than immediate treatment and relies on capacity elsewhere in an accessible network. In reality in a warfighting context where local infrastructure is damaged beyond immediate repair, this may not be feasible and so alternative solutions will be required.

The NHS has had to prioritise learning how to find alternatives to admissions into an overcapacity system over the last decade, and from this work have come concepts such as ambulatory care, patient-delivered treatments and community outreach teams. All of these concepts may have a role in a military MASCAL situation, reducing the requirement for admission and load on inpatient facilities, but all of these processes have to be designed beforehand. In this scenario, those who are ill or injured but are unlikely to come to significant harm without hands-on medical treatment could be sent home or, if able, returned to some form of duty.

This process could be enhanced by the use of pre-prepared self-care packages. For example, for a patient with a broken wrist a thermoplastic splint and simple analgesia could be administered, allowing them to manage away from the MTF until such time as there was sufficient capacity to provide more definitive care. Another way of enhancing this 'care in the community' equilvalent approach is to give patients wearable physiological monitors to take with them to give an early warning of deterioration. This again relates to NHS experience. During the coronavirus (COVID-19) pandemic, NHS patients were successfully discharged home and managed remotely with the help of pulse oximetry monitoring (Shah et al. 2020). In addition to the ability to measure traditional physiological parameters such as heart rate and rhythm or oxygen saturations, more advanced sensors can detect other signs of impending deterioration. For example, measuring the amount of cortisol in sweat can provide an early warning of increasing sympathetic drive due to bleeding or other causes of deterioration.

This kind of discharge decision is, however, extremely challenging. In usual UK emergency medicine (EM) practice, even doctors with six- or seven-years post-graduate experience will often feel the need to discuss discharge of patients with a consultant emergency physician. In the exceptionally challenging contest of a MASCAL there is an argument for ensuring a senior clinician is involved in the decision-making. If this is viewed with the potential requirement for a senior clinician to undertake anatomical, non-batched triage there is a significant demand for senior clinicians who will already be being pulled in many different directions. This may not be possible. In which case there may have to be an acceptance that mistakes will be made by more junior clinicians, who may not be doctors, who are being asked to make very difficult decisions, in very short time frames in a challenging and emotionally fraught environment. We suggest that in order to allow people to make the best decision for the majority there needs to be an explicit acceptance of the risk of error, both legally and ethically. In the case of non-doctors being asked to make these decisions there may also be a need for an explicit acceptance of a change in scope of practice from regulatory bodies, for example, the Nursing and Midwifery Council or Health and Care Professionals Council in this specific set of circumstances. Again this requires deliberate planning before the event. We will be returing to these issues in Sect. 3.5.

We will now explore our second case study, medical provision on the Western Front during the First World War. We consider that Britain's First World War is an effective comparison point and model for the immediate and developing response process to a mass casualty event or disaster. This war was a mass casualty event lasting almost five years, and provides a site from which to study both response and resource processes as they are developed and constrained by both time and the pressure of numbers.

3.3 Case Study 2 – Historical Considerations: Mass Casualty Management on the Western Front of the Great War (1915–1918)

By early 1915, casualty and disease numbers in the British Army generated by the fighting on the European Western Front had crossed the threshold of functionality for the existing medical system. Phases of attritional war fighting produced waves of mass casualties suffering explosive and chemical weapons trauma, bacterial infection and disease outbreaks across broad land-based battlefronts. Commonalities between then and now are significant. Whether volunteer or conscript, this was a young, male, fit population that had benefited from innovations in public health, particularly systems of inoculation, improved sanitation and the disciplined implementation of the sanitation regulations by the Regimental Medical Officer at battalion level. Infection management in the pre-antibiotic era was prioritised but limited, and it is likely this will be seen as a forerunner of our own antibiotic-constrained times. The most significant difference across the century was in the expectation of standards of care. There was no NHS, and no understanding of emergency or trauma care, or of an equivalent of 'gold standard care' by either practitioner or patients. Only those who previously worked in heavy and extractive industries were accustomed to organised teams of first responders providing rescue and on-site care before transportation to company or municipal medical facilities.

These occupational first responders would provide the model for the reorganisation of military medical provision in the mass casualty situations of the Western Front. Capability was moved as far forward as it could go, as close to the point and time of wounding as possible. A new medical cadre was created to deliver this capability: the battalion stretcher bearers reporting to the Regimental Medical Officer as a team. Control of haemorrhage was a priority: British stretcher bearers were the first cadre without a medical degree to be formally trained to manage wounds, both at the point of wounding and along the casualty evacuation chain. Beyond technical first responder skills, history shows us that bearer teams were also empowered to make significant decisions throughout the process, and that their unique expertise and experience was recognised for its absolute value in securing the flow of casualties through the reoriented field medical sites. Bearers chose when their casualties were strong enough to be moved, and when it was safe enough for movement to take place. They allocated priority and recognised futility and its consequences. They chose their evacuation route along multiple aid post sites that they had pre-planned and pre-supplied. They managed the pain of their patient within the context of the expectation of immediate damage control surgery so that anaesthetics could be quickly administered without waiting for the effects of other medication to wear off.

At the heart of bearer expertise was the understanding of the flow of their patients along the casualty pathway that ran from aid post to clearing station to field hospital. They constantly tested the routes they had selected, and monitored them for damage from weather or nearby fighting. They learned new roads into field hospitals that took their patients straight to the resus tent. They formalised the physical and administrative processes of entraining and detraining patients on to the hospital train network, creating specialist teams who did nothing else, and could load a full train in almost complete darkness and at pace. They enabled the creation of a structure to manage the dead, preparing the moribund wards and death tents and temporary cemeteries before battles began. Their priority for this practice reminds us of the need for advanced planning for management of the dead within the mass casualty process, not only for the families of patients but also for the medical staff of whatever cadre to maintain their ongoing morale.

No matter what capacity, capability and decision-making expertise this new medical cadre was given in the Great War, the system could be overwhelmed. In 1916 the battle of the Somme saw field hospitals either swamped with patients inside and outside their huge tented wards, or lying empty because the rail and road networks that linked them to the battlefield were destroyed. Bearers found themselves stuck in no man's land with their patients, unable to move and no longer able to identify where their destination was because of the huge explosions that wrecked the landscape. Using whatever supplies were to hand, they created and ran temporary aid posts, adapting their capabilities to sustaining the lives of their patients until flow could resume. At field hospitals, they supported the mass of casualties unable to receive immediate medical care, and facilitated self-care and care by comrades with distributions of packages of water, dressings and analgesics.

The battle of the Somme was the last time British military medical provision on the Western Front was significantly overwhelmed. Empowering the lowest capable provider for both clinical and non-clinical decision-making and operations had become the standard. By the end of the battle, bearers had also refined their own particular skillsets. They were now experts in the casualty of evacuation of patients with severe femoral fracture, by combining bespoke splinting with careful carriage (CASEVAC as we would term it now). Death rates from femoral fracture fell from 60% to 10% by the end of 1916. Despite the rapid development and consolidation of this new, multi-disciplinary expertise, the military medical system failed to take advantage of the learning it represented. It was not deemed suitable to record the achievements of these lower capability providers no matter how many lives their skills had secured.

We believe that appropriately empowering the lowest capable provider (LCP) is an important part of the solution to managing MASCAL incidents. Therefore, we will now explore this in more depth. We will also explore the crucial change in mindset that is required to successfully manage a MASCAL incident; a shift from providing gold standard care to providing the most appropraite care in the circumstances.

3.3.1 Lowest Capable Provider

Both historical and recent experiences show that empowering first responders saves lives. The introduction of tourniquets during the conflicts in Iraq and Afghanistan had an enormous impact on deaths from massive limb haemorrhage. These injuries had accounted for 7.5% of battlefield fatalities. After these tourniquets entered widespread service deaths from these injuries dropped to 2.6% (Kotwal et al. 2013). Crucially, these lives were saved not by doctors, but by soldiers. This trend to 'front-loading' care by empowering lower-level providers continues. Recently administration of Tranexamic acid (a drug to reduce bleeding in trauma) has been rolled out to the front lines in auto-injection devices designed to be self- or buddy-administered by soldiers (Gov.uk. 2019). In the context of MASCAL, a driving principle should be that, whenever possible, care should be given by the lowest level of capable provider. Accepting that this brings enormous benefits in this context sits well with the military drive to push care to the point of wounding. Logically the use of innovation to support one will likely assist the other and so adopting the principle gives an effective focus of technological research for both areas.

3.3.2 Context Specific Excellence: The 'Best Care Feasible' Vs 'Gold Standard Care'

Clearly some of the suggestions made thus far have implications for the care that is provided and these implications need to be addressed head on. There is a tendancy to assume that there is a defined best option care for all non-novel eventualities. Medical knowledge is, however, constantly evolving and so the gold standard today will be 'better' in some demonstrable fashion than the gold standard of yesterday. These improvements usually come, however, at a cost in terms of resources, and usually follow a law of diminishing returns. In other words, we use increasing resource expenditure for smaller and smaller incremental gains. Such incremental improvements are not obviously defensible in the resource critical environment of a MASCAL.

An example is the care of a specific, dangerous heart attack, known as an 'ST elevation Myocardial Infarction' (STEMI). For 20 years around the millennium, the clot causing a STEMI was 'dissolved' using thrombolytic drugs. These drugs were expensive (£700 per patient in the UK) but could be administered as a bolus by a competent EM doctor, paramedic or physician. This practice has been replaced by mechanical displacement of the clot using a stent, inserted endovascularly through a peripheral artery. The 'average' STEMI now costs $19,000 for initial hospital stay (Cowper et al. 2019), and often requires a cardiac catheter laboratory with a consultant cardiologist supported by a nurse, physiological assistant, radiographer and potentially an anaesthetic team of an anaesthetist and ODP/anaesthetic nurse. The absolute survival benefit of the new treatment at 1 year is in the region of 2.7% – with mortality at 7.6% after a stent, and 10.3% after thrombolysis (Stenestrand et al. 2006). In essence, while the old gold standard is not as good as the new one, it is still good, leading to a much lower mortality than STEMIs that are untreated. Delay to treatment of a STEMI is a key predictor of outcome, so in most economically advantaged countries the target is for the clot to be treated in under two hours (Scholz et al. 2018). When there is a long delay, the benefit of stent over thrombolysis fades. In a MASCAL situation, timely administration of a drug given by one doctor with a syringe is much more feasible than treatment by a team of up to six in a dedicated theatre. Perhaps in a MASCAL a 'silver' standard may be as good, or even contextually better than attempting to apply the current 'gold' standard.

Critical to this concept is that fact that you cannot suddenly revert to a standard that was the norm ten years ago in the middle of a MASCAL. There is a danger that those used to delivering the most uptodate care will not remember how to use or perform quicker, less resource intensive methods that are no longer routinely used. Moreover, the relevant drugs or equipment may no longer be immediately available. In orer to deliver the most appropriate standard of care in the MASCAL context, you have to plan for it.

Once freed to consider the option of a 'context-specific gold standard', or *what excellence looks like in that situation*, it is possible to look objectively at a range of likely conditions to see what treatments are critical, and which are simply desirable. For example, for a patient with a broken arm, most emergency physicians would consider arranging an Xray as part of the gold standard. But if the arm is clearly bent, and no intervention is possible right now because of resource constraints, is it really helpful? No. Most EM physicians in a disaster context would apply a splint for comfort, give analgesia, and discharge the patient for later follow up. The Xray can be safely deferred, arguably, in many cases, omitted. Given an ED takes at least 11 min to perform an Xray, omitting some Xrays in a safe, planned fashion would significantly reduce treatment times (Kwok et al. 2021).

3.3.3 Focussing on Quality End of Life Care

Little research has been done into what futility means for trauma patients, and in the military it has been considered especially problematic. Given that the casualties have come from a predominantly fit and healthy young population, in a gold standard system there have been successful outcomes despite injuries previously considered "unsurvivable" (Russell et al. 2011). However, the outstanding results delivered by the medical services during the conflict in Afghanistan came at a heavy cost in terms of resource and logistics. In warfighting, the level of resourcing available would not be close to that seen in Afghanistan, and so similar outcomes are unlikely to be deliverable. Particularly in a MASCAL situation there will have to be rationing of resource both in terms of physical treatments and also time.

Similar 'competition' for scare resources is seen in the civilian sector in serious pandemics, where critical care capacity is overwhelmed and gold standard care cannot be delivered to all who need it. For example, the NHS has accepted that triage may be applied to exclude those who are least likely to recover. In the 2009 pandemic influenza plans, examples of exclusions were severe trauma, severe burns, advanced malignancy, pre-existing severe cardiac disease, a SOFA score >11 or lung disease with a FEV < 25% (Department of Health (UK) 2009). A similar proposal was made in the light of the coronavirus (COVID-19) pandemic (Cook et al. 2020). Given that in a grossly overloaded system, accepting standards below those normally provided is a pre-existing concept, military medical services should aspire to support the decision-making of their clinicians by trying to understand who is most likely to benefit from available resources in a MASCAL context, in advance of the event. A benefit of this approach is that it also gives permission to explore how those casualties where active treatment would be deemed futile (given the resource constraints) can be most humanely managed. This would make 'dying well' a specific care objective – rather than pointless overtreatment or simple neglect that they would be likely to receive in a MASCAL context currently. What is the best way to provide dignity and comfort in death to large numbers of catastrophically injured patients? In World War 2 the answer often involved a priest and a cigarette. Perhaps the padre (or other non-medical support workers) can re-appraise this role, but with a greater armamentarium at their disposal. Could they, for example, in the very limited context of a MASCAL be allowed to administer drugs that would manage the pain and distress of the dying?

3.4 Some Ethical Considerations

In this section, we will highlight and discuss some of the ethical dimensions of the proposals we have outlined in the preceeding sections, starting with the proposal that the standard of care should be 'the best feasible'.

There seems to be general reluctance to introduce into policy anything less than gold standard care, even in the context of a MASCAL on the scale of the incidents outlined above, or in the context of overwhelmed hospital services during a pandemic, as experienced at times in the UK NHS during 2020–2021 with the coronavirus (COVID-19) pandemic. In the context of policies for outside/expat responders to natural disaster – such as the Haiti earthquake – such reluctance makes a little more sense. Giving a green light to lower standards in disasters with mass casualties could lead to a medical 'have a go' free-for-all, with irreversibly poorer outcomes, justified by a potentially false belief that something is better than nothing. For example, the amputation of severely injured limbs that could have been salvaged by a more conservative approach. This is one of the reasons why funding those agencies with expertise in, and considerable experience of, disaster management is the appropriate response to disasters overseas: it minimises the harms caused by those with good intentions but none of the requisite skills and knowledge. In normal times, in an organisation like the NHS where funding does not keep pace with public expectations, it is inevitable that in some already pressured departments during peak periods (like winter pressures) not all patients receive gold standard care. Many do. Some receive 'good enough' care and some fall between the cracks with tragic results. Of course, it is not right that some patients receive less than optimal care in a system that promises equity and equality. Equally, the pressure on staff to deliver the "highest standards of excellence and professionalism" (Department of Health (UK) 2021) is unfair if they are not provided with the resources to deliver these standards. The NHS is a public institution, funded from taxation. It exists in a constant state of tension between meeting need and delivering efficiency. In many respects, this tension can be a productive one that keeps waste to a minimum and which is constantly exploring ways in which care can be delivered as cost-effectively as possible. But there's the rub. The metrics employed to determine cost-effectiveness are not designed to provide "excellence" to each and every patient. For example, the National Institute for Health and Care Excellence (NICE – the body in the UK charged with providing evidence-based guidance to improve the health outcomes of NHS service users) states:

> Generally, we consider that interventions costing the NHS less than £20,000 per QALY [Quality Adjusted Life-Year] gained are cost effective. Those costing between £20,000 and £30,000 per QALY gained may also be deemed cost effective, if certain conditions are satisfied (National Institute for Health and Care Excellence 2013).

Although NICE may not reject an intervention on the grounds of cost alone, cost-effectiveness is assessed against the next best alternative. In short, less good alternatives might offer better value for money even though they are not optimal. Very expensive therapies may be rejected if the benefit derived from them (as assessed by QALYs, not by the patient or the patient's doctor) is deemed insufficient to justify the cost. This is not to pick a fight with NICE, or even the notion of providing the best value for money in a welfare state. Rather, it demonstrates that terms like 'excellence' and 'gold standard' seems at odds with reality both on the ground and in funding policy and can therefore be questioned. Resources are rarely sufficient to meet demand – patients may wait for months for hospital appointments, diagnostic tests and then for surgery, for example. Whilst most people usually do eventually get their 'non-urgent' surgery, waiting times are certainly not compatible with optimal care.[1] Moreover, the so-called gold standard isn't actually gold at all (unless 'gold' is not intended to imply 'best' – perhaps platinum care would be better still?): policies for cost-effectiveness and waiting lists attest to this. Notions of 'excellence' and 'gold standard' are political, rhetorical tools that are used because promising

[1] It is possible to imagine circumstances where building in waiting time may be regarded as advantageous to autonomous decision-making.

something that is 'good enough' is, well, not good enough to win votes and inspire confidence. But use of such terms without clear qualification is dishonest. At the best of times, it puts individual NHS staff members under huge pressure, and, as we have suggested, makes it very difficult to make realistic plans for truly dreadful circumstances. We set our medical staff up for inevitable failure if we do not openly accept that standards of care *in extremis* will change, and this is, quite simply, unfair.

So, can we still expect excellence from staff working in MASCAL events? In short, the answer to this question is 'Yes', but only if what we expect is people, in good faith, to strive for excellence in the Aristotelian sense of 'functioning well', because in this sense excellence is relative to circumstances, as well as being a habit or character trait. An excellent plan for an overwhelming MASCAL event will be one that achieves the best that can be achieved in those circumstances. It might require extraordinary efforts – as befits extraordinary circumstances – but it will not be set up to fail. Doctors and other healthcare professionals should not need to be reassured that their actions will be judged against the prevailing, very challenging circumstances (General Medical Council 2020). This should be obvious from the outset, and it is a failure of existing guidance and communication of this guidance that this is not already clear.

The UK Government's pandemic planning contained an Ethical Framework (Cabinet Office 2013), which was amended in the light of the WHO's ethical guidance for infectious disease outbreaks (World Health Organization 2016). In its guidance the UK government documents states:

> Equal concern and respect is the fundamental principle that underpins the ethical framework. This means that:
>
> - everyone matters
> - everyone matters equally – but this does not mean that everyone is treated the same
> - the interests of each person are the concern of all of us, and of society
> - the harm that might be suffered by every person matters, and so minimising the harm that a pandemic might cause is a central concern

At a time of acute crisis, minimising the harm may be the best that can be achieved. Recognising the need to minimise harm is compatible with accepting that some people will experience harm nevertheless. In this respect it may also be important to draw a distinction between being harmed and being wrong. Harms befall us all of the time. Sometimes harm results from misfortune and sometimes from our own choices. By and large we cannot really complain about these sorts of harm.[2] Harms become wrongs when they are unjustly inflicted, which is in part why it is important to stress that "everyone matters" and "everyone matters equally".

Planning itself can contribute to harm minimisation. Provided the plan fits the crisis, it may lessen early, harmful disarray. Planning also facilitates public and other stakeholder consultation, which: reduces the risks of harms being unfairly

[2] Unless we are more likely to be unfortunate or make choices that increase the prospects of harm because of some systemic injustice. An example here might be the health disparities that result from poverty, which itself results from social biases that make it harder for some individuals to become financially secure than others.

allocated as a result of unconscious bias, increases the prospects of all available alternatives being explored, and increases trust and aids resilience. All this is only possible if those plans include unpalatable truths. One such truth is that resources will be insufficient, people are going to die who might have survived in circumstances where more resources are available, and that planning to make the best possible use of the resources that are available is a way of minimising harm, and this includes lowering the standard of care. To some extent it really doesn't matter how this is phrased: 'good enough', 'silver standard', or 'best care feasible' etc. What it should probably not be called, though, is 'suboptimal', precisely because it *is* what is optimal under the circumstances and we must not suggest to those who have to operationalise the plans that they have failed even before they have started.

What may be lacking in the military is stakeholder consultation. This is not to say that British military policy is not subject to external scrutiny and influence – it is certainly the object of political scrutiny and sensitivity. Perhaps engagement with those who might be most affected – the troops on the ground and those providing the care – is not something that the military feel able to continence. It might be thought to undermine morale or the chain of command. The former, however, may be a matter of timing. Leaving engagement until the eve of battle would be bad timing but also bad planning – plans should be in place well before this. Rather than undermining morale, involvement in planning might improve confidence and understanding. After all, given that self-care is likely to be significant component of the plan, individuals will have to be trained so they do have to know what the plans are, why they are in place and what the stakes will be in a MASCAL.

One of the potential advantages of military planning for a military-on-military scenario where "casualties in the hundreds or even thousands per day could be managed, especially if this were to continue over days or weeks" (as described above) is that the bulk of the population at risk will be fairly harmonious: fit, youngish combatants. Although no one will know precisely what their position will be when the veil of ignorance, as it were, is lifted, the possibilities will tend to be limited to the extent of their injuries. Thus, it is reasonable for them to be assured that the plans in place will maximise their prospects of survival and minimise their prospects of long-term morbidity, in the context of the situation in which they treated. The risks of being exposed to injury, whether or not they are injured, and the extent of those injuries, is down to strategic factors that are understood to be part of their chosen professional role,[3] and some degree of luck. We might not, therefore, have to be *as* concerned about whether existing socio-economic disadvantage will be compounded by the response to a military MASCAL event, provided that purely clinical indicators are used to determine whether individuals are treated, made comfortable, 'discharged' or transferred. Disadvantage may be compounded, on the other hand, if apparent ethnicity, gender, rank or affiliation to a particular unit influences the choices made, and it would be unwise to discount unconscious bias at work in practice, even if this is not reflected in policy.

[3] This point assumes that the personnel in question are not conscripts.

Finally, we should consider how care as well as clinical decision-making and treatment should be factored into planning for these kinds of events.

What is meant by 'care' in this context? While we can all identify colleagues who fail in this regard, by and large when we talk about providing 'care' in a medical context we are referring to both the intervention and the way in which this is delivered and we do not normal distinguish between these two elements. 'Care' is an interesting word, the noun 'care' (that which is given by the professional and received by the patient) incorporates both the verb and adverb ('to care' 'caringly'). Sometimes 'care' is being delivered, but not perceived as such. For example, when the staff are wearing extensive personal protective equipment, or the patient is unconscious. Virtuous and vicious cycles can be set up around care. A virtuous cycle develops when care is delivered and acknowledged as such. A vicious cycle can emerge when care is being delivered but cannot be perceived: it takes more of an effort of both professional and personal will to continue 'to care' in these circumstances, and not to become disinterested or callous in the delivery of interventions. Where this occurs, the two elements that make up care as normally understood, have become uncoupled; an intervention is delivered but this is not 'care' as normally understood.

Next, a caveat: the value of care is relative. Normally, we might claim that care is extremely valuable. It may even have a definitive role in health service provision. In abnormal times, such as a massive and overwhelming MASCAL event, the main goal of preserving life and limiting morbidity for as many as possible, may displace other values. We can see this, for example, in relation to respect for patient autonomy, another core value. When personnel and resources are very stretched, it may be justifiable to curtail respect for patient autonomy, particularly when the driving justification for decision-making is overall population benefit. The time taken to make a patient a partner in their own treatment is time taken away from providing treatment to another person. Catering to individual preferences about treatment may undermine their effectiveness. At the peak of the waves of the coronavirus (COVID-19) pandemic, for an example, it is arguable whether a patient who was offered ventilation when ICU beds are at a premium, should be allowed to negotiate for more time on continuous positive airway pressure therapy (perhaps to say goodbye to a loved one), when the upshot of the delay is that ventilation might be less successful. Of course, the patient might be entitled to refuse ventilation, but this refusal cannot be backed-up with a guarantee that in even a few hours' time, the offer of ventilation is still on the table. In other words, in times of scarcity the offer is in effect a 'take it or leave it' one, not a negotiation. We can debate whether this choice, harsh as it is, is coercive or not given the constraints. The purpose of the example is to make the point that respect for patient autonomy normally translates into accommodating preferences in ways that, at a time of crisis or emergency, might not be possible. In a MASCAL event, it may be reduced to merely telling (rather than explaining to) someone what you are doing, with no choice available: 'I am giving you A because B and then C will happen.' Just as the norm of respect for patient autonomy may be suspended, so too might aspects of the delivery of interventions that are associated with care. There may be no time for moments of sustained eye contact, or touch, or comfort and reassurance when there are huge numbers to assess and treat. Indeed, the situation may be so overwhelming that the staff go into 'automatic' mode. We can but hope that some element of care is contained in their 'default' such that they are at least not callous, as they make one apparently brutal decision after another, for days at a time.

So, can we make any provision for 'care' in plans for MASCAL events? We offer three suggestions for how this might be done. The first leads directly from the previous paragraph: we need to ensure that providing care is so ingrained in staff that at some level it survives in their default or automatic mode. But, moving to the second suggestion, the effect this may have on staff also needs to be factored in. Not being able to provide care when this is part and parcel of who you are is psychologically damaging for staff: we have seen this in the mounting mental health crisis amongst NHS staff over the coronavirus pandemic (British Medical Association 2021). Plans must not set staff up for failure, so the plan itself needs to factor in how to prepare them for the awfulness of MASCAL events on the scale we are discussing. This means learning as much as we can about how to protect the mental wellbeing of staff from events and circumstances where we have previously failed to do so, and working out what measures can be put in place to lessen this. We must not forget that whilst healthcare professionals will rise to the most enormous challenges, and we might even argue that they are a duty to do so (Draper 2021), they have no duty or obligation to remain within the employ of either health or military services once the crisis is over. So, we may survive one catastrophe only to be in a worse position to meet the next.[4]

The third way in which care could be factored into MASCAL plans concerns those for whom treatment is likely to be futile: the ones who are going to die, may be sooner, may be a little later but who are not going to be offered treatment. They have had the mucky end of two sticks: they have sustained massive injuries and they will be passed over by medical responders *because* their injuries are so great. This must be acknowledged. It should be possible to build into MASCAL plans *something* for the dying. Here again, we can consider the lowest capable provider, which need not be one of the medical responders. The military have some history of using a variety of staff – the padre for example – and even other patients to offer comfort to the dying, for example touch or holding a hand, receiving last messages for loved ones, lighting and holding a cigarette to their lips, and hopefully administering pain relief if this is available. Including some provision for the dying may be reassuring to those about whom these plans are being made (those who may at some point be one of the casualties in a MASCAL). It may also help medical responders to perform the difficult task of treating selectively if they know that someone else has the task of caring for those they will inevitably have to step over or roll to one side in order to treat their next patient.

---

[4]There are person-centred reasons as well as consequential reasons for protecting staff: it is not obvious, however, that political planners care as much about individuals as they do about wider consequences.

3.5 Summary

Mass casualty incidents overwhelm services in a way that is completely different from major incidents, but as yet most guidelines do not recognise the implications of this. Given the strong evidence that gold standards of care are not always maintained in pressured healthcare systems such as the NHS, and that the associated overcrowding results in harm, it must be assumed that gold standards will fail in MASCAL events. Historical lessons from the World Wars, as well as current coping strategies employed by the NHS give insights into how to minimise the harm incurred. A key strategy may be to plan for a context dependent standard of excellence, different from day-to-day care, that allows a focus on delivering best possible quality with limited resources. A range of ethical reasons to consider this approach, rather than simply expecting providers to improvise at the time, have been explored.

References

Abdul Nabi, S.S., and R.D. Sawaya. 2020. Airway breathing circulation: An emergency medicine resident's experience of the Beirut explosion. *Academic Emergency Medicine* 28 (4): 483–486.

Advanced Life Support Group. 2011. *Major incident medical management and support: The practical approach at the scene*. Wiley-Blackwell.

BBC news. 2020. *Beirut explosion: What we know so far*. Available from: https://www.bbc.co.uk/news/world-middle-east-53668493

Bricknell, M.C.M. 2003. The evolution of casualty evacuation in the British Army in the 20th century (Part 3)–1945 to present. *Journal of the Royal Army Medical Corps.* 149 (1): 85–95.

British Medical Association. 2021. *Year of unparalleled pressure leaves staff exhausted and on the brink, BMA survey shows*. Available from: https://www.bma.org.uk/bma-media-centre/year-of-unparalleled-pressure-leaves-staff-exhausted-and-on-the-brink-bma-survey-shows Accessed 8 July 2021.

Cabinet Office, Department of Health and Social Care, Ministry of Housing CaLG, Home Office, Justice Mo. Pandemic flu: pandemic flu planning information for England and the devolved administrations, including guidance for organisations and businesses. 2013. (Updated 2017).

Calver, T., and Lintern, S. 2022. Deaths rise as A&E crisis leads to ambulance delays. *The Times*. 21/8/2022.

Cook, T., K. Gupta, C. Dyer, R. Fackrell, S. Wexler, H. Boyes, et al. 2020. Development of a structured process for fair allocation of critical care resources in the setting of insufficient capacity: A discussion paper. *Journal of Medical Ethics*.

Cowper, P.A., J.D. Knight, L. Davidson-Ray, E.D. Peterson, T.Y. Wang, D.B. Mark, et al. 2019. Acute and 1-year hospitalization costs for acute myocardial infarction treated with percutaneous coronary intervention: Results from the TRANSLATE-ACS registry. *Journal of the American Heart Association*. 8 (8): e011322.

Department of Health (UK). 2009. *Pandemic flu: Managing demand and capacity in health care organisations* (Surge).

———. 2021. *The NHS constitution for England*.

Draper, H. 2021. Risk and infectious disease outbreaks: Should military medical personnel be willing to accept greater risks than civilian medical workers? In *Health Care in Contexts of risk, uncertainty and hybridity*, ed. D. Messelken and D. Winkler. Cham: Springer Nature.

Emergency Preparedness RaR. 2017. Concept of operations for managing mass casualties. In: England N, editor.

Ewbank, L., J. Thompson, and H. McKenna. *NHS hospital bed numbers: Past, present, future: The King's fund; 2017*. Updated September 2017. Available from: https://www.kingsfund.org.uk/publications/nhs-hospital-bed-numbers

General Medical Council. 2020. *Joint statement: Supporting doctors in the event of a COVID-19 epidemic in the UK*. Updated 11 March 2020. Available from: https://www.gmc-uk.org/news/news-archive/supporting-doctors-in-the-event-of-a-covid19-epidemic-in-the-uk.

Gov.uk. 2019. *Life saving frontline technology given £5-million boost*. Available from: https://www.gov.uk/government/news/life-saving-frontline-technology-given-5-million-boost

Home Office. Coroner's inquests into the London Bombings of 7 July 2005 2012.

Joint emergency services interoperability principles (JESIP). 2016. Joint doctrine: The interoperability framework. JESIP. 2/6/2016.

Kipnis, K. 2003. Overwhelming casualties: Medical ethics in a time of terror. *Accountability in Research: Policies and Quality Assurance*. 10 (1): 57–68.

Kotwal, R.S., F. K. Butler, E. P. Edgar, S. A. Shackelford, D. R. Bennett, and J.A. Bailey. 2013. *Saving lives on the battlefield: A joint trauma system review of pre-hospital trauma care in combined joint operating area-Afghanistan* (CJOA-A).

Kwok, M.M.K., A. Chiu, J. Chia, and C. Hansen. 2021. Reducing time to X-ray in emergency department ambulatory patients: A quality improvement project. *BMJ Open Quality* 10 (2): e000995.

Montán, K., A. Khorram-Manesh, and P. Ortenwall. 2011. Comparative study of physiological and anatomical triage in major incidents using a new simulation model. *American Journal of Disaster Medicine* 6: 289–298.

National Institute for Health and Care Excellence How NICE measures value for money in relation to public health interventions. 2013.

NHS interim management and support. 2011. *Effective approaches in urgent and emergency care*.

North Atlantic Treaty Organisation (NATO). 2015. *Medical aspects in the management of a major incident/mass casualty situations*.

Pines, J.M., C.V. Pollack Jr., D.B. Diercks, A.M. Chang, F.S. Shofer, and J.E. Hollander. 2009. The association between emergency department crowding and adverse cardiovascular outcomes in patients with chest pain. *Academic Emergency Medicine*. 16 (7): 617–625.

Reminick, G. 2001. *Nightmare in Bari. The world war II liberty ship poison gas disaster and cover up*. The Glencannon Press.

Russell, R., T. Hodgetts, J. McLeod, K. Starkey, P. Mahoney, K. Harrison, et al. 2011. The role of trauma scoring in developing trauma clinical governance in the Defence medical services. *Philosophical Transactions of the Royal Society B: Biological Sciences*. 366 (1562): 171–191.

Scholz, K.H., S.K. Maier, L.S. Maier, B. Lengenfelder, C. Jacobshagen, J. Jung, et al. 2018. Impact of treatment delay on mortality in ST-segment elevation myocardial infarction (STEMI) patients presenting with and without haemodynamic instability: Results from the German prospective, multicentre FITT-STEMI trial. *European Heart Journal*. 39 (13): 1065–1074.

Shah, S., K. Majmudar, A. Stein, N. Gupta, S. Suppes, M. Karamanis, et al. 2020. Novel use of home pulse oximetry monitoring in COVID-19 patients discharged from the emergency department identifies need for hospitalization. *Academic Emergency Medicine*. 27: 681–692.

Stenestrand, U., J. Lindbäck, and L. Wallentin. 2006. Long-term outcome of primary percutaneous coronary intervention vs prehospital and in-hospital thrombolysis for patients with ST-elevation myocardial infarction. *Journal of the American Medical Association* 296 (14): 1749–1756.

Sun, B.C., R.Y. Hsia, R.E. Weiss, D. Zingmond, L.-J. Liang, W. Han, et al. 2013. Effect of emergency department crowding on outcomes of admitted patients. *Annals of Emergency Medicine*. 61 (6): 605–611.e6.

The History Press. 2021. *Evacuation of the wounded in World War I*. Available from: https://www.thehistorypress.co.uk/articles/evacuation-of-the-wounded-in-world-war-i/

Watling, J. 2019. *The future of fires. Maximising the UK's tactical and operational firepower*. Royal United Services Institute for Defence and Security Studies; November 2019.
Wikipedia contributors. 2004. *Madrid train bombings: Wikipedia, The Free Encyclopedia*. Updated 14/7/21. Available from: https://en.wikipedia.org/wiki/2004_Madrid_train_bombings
———. 2017. *Las Vegas shooting: Wikipedia, The Free Encyclopedia*; 2021.
———. 2021a. *Bhopal disaster: Wikipedia, The Free Encyclopedia*. Available from: https://en.wikipedia.org/wiki/Bhopal_disaster
———. 2021b. *San Juanico disaster: Wikipedia, The Free Encyclopedia*. Available from: https://en.wikipedia.org/wiki/San_Juanico_disaster
World Health Organization. 2016. *Guidance for managing ethical issues in infectious disease outbreaks*.

Col Simon Horne is an Emergency Medicine Physician with research interests in Major Incident management and triage. His deployments include Iraq and Afghanistan, as the medical director of the UK Military Ebola treatment unit in Sierra Leone and also with the UN to South Sudan. He is head of the UK Centre for Defence Healthcare Engagement.

Wing Commander Robert James is a Consultant in Emergency Medicine and Pre-Hospital Emergency Medicine. He is an Honorary Senior Lecturer in Military Emergency Medicine for the Academic Department of Military Emergency Medicine and an Honorary Lecturer in Pre-Hospital, Retrieval and Transfer Medicine for the University of Plymouth. His research focuses on the resuscitation of injured patients; he has published papers and book chapters on this topic.

Heather Draper is Professor of Bioethics at the University of Warwick. She has published widely in bioethics including on military medical ethics issues.

Emily Mayhew is a military medical historian specialising in the study of severe casualty, its infliction, treatment, and long-term outcomes in twentieth and twenty-first century warfare. She is historian in residence in the Department of Bioengineering, working primarily with the researchers and staff of the Centre for Blast Injury Studies. She is also a Research Fellow in the Division of Surgery within the Department of Surgery and Cancer. She is based jointly in the Department of Bioengineering and at the Chelsea and Westminster campus. She is Imperial Internal Lead on the Paediatric Blast Injury Partnership.

Chapter 4
Medical Rules of Eligibility: A Comparative Analysis

Joanne Clifford, Paul C. Eagan, and Sheena M. Eagan

Abstract International Humanitarian Law/Law of Armed Conflict requires that medical care be provided without distinction. The realities of armed conflict, however, have complicated this universal rule. Specifically, resource scarcity within the context of armed conflict has led to medical policy and practices that seem to contradict the foundational concept of non-discrimination. Simply put—sometimes health care providers cannot provide care to everyone, necessitating distinction between patient groups based not only upon medical need, but also military necessity, and patient identity (e.g., compatriot, ally, enemy, civilian). Policies which define an individual's eligibility and the level of medical care that they will receive from deployed medical treatment facilities are known as medical rules of eligibility or MROE. This chapter offers a comparative analysis of established policies surrounding MROE within NATO, the Canadian Armed Forces (CAF), and the United States Department of Defense (DoD) medical doctrine, providing an overview of how medical eligibility decisions are formalized into policy across military contexts.

Keywords Medical rules of eligibility · Military health care · Medical eligibility decisions

J. Clifford (✉)
Defence Ethics Programme, Department of National Defence, Ottawa, ON, Canada

P. C. Eagan
Canadian Forces Health Services, Ottawa, ON, Canada

Dalhousie University Faculty of Medicine, Halifax, NS, Canada
e-mail: paul.eagan@forces.gc.ca

S. M. Eagan
Department of Bioethics & Interdisciplinary Studies, Brody School of Medicine, East Carolina University, Greenville, NC, USA
e-mail: eagansh17@ecu.edu

S. M. Eagan, D. Messelken (eds.), *Resource Scarcity in Austere Environments*, Military and Humanitarian Health Ethics,
https://doi.org/10.1007/978-3-031-29059-6_4

4.1 Introduction

It is a core tenet of International Humanitarian Law/Law of Armed Conflict (IHL/LoAC) that medical care in military operations be provided without distinction.[1] The realities of armed conflict, however, often serve to complicate this universal rule. Resource scarcity, particularly within the context of armed conflict, has led to policy and practice that seem to contradict the foundational concept of non-discrimination. As a result, healthcare providers may find themselves in the unenviable position of not being able to provide care to all of those who need it, necessitating distinction between patient groups based upon not only medical need, but also military necessity and patient identity, for example, compatriots, coalition forces, allies, enemies, civilians. In the military environment, the policies that define an individual's eligibility for care, as well as the level of medical care that they are entitled to receive at deployed military medical treatment facilities, are referred to as Medical Rules of Eligibility or MROE.

In this chapter we offer a comparative analysis of military MROE within NATO doctrine and that of two member nations: Canada and the U.S. Our intent is not to critique or to evaluate the doctrine, but rather to introduce the guiding doctrine and provide insight into how medical eligibility decisions are formalized into policy and doctrine across NATO, the Canadian Armed Forces (CAF) and the United States Department of Defense (DoD). In addition, we compare the ease of access to policy, as well as identify some of the key doctrinal gaps and differences.

Thus, this comparative analysis begins with an exploration of two key NATO publications: NATO Standard Allied Joint Publication 4.10 (AJP-4.10): *Allied Joint Doctrine for Medical Support* (September 2019), and its subordinate publication, NATO Standard Allied Joint Medical Publication 1 (AJMedP-1): *Allied Joint Medical Planning Doctrine* (September 2018). Subsequently, two CAF publications are reviewed. The first is the *Canadian Forces Joint Publication 4-10, Health Services* (CFJP 4-10) an interim publication released in September 2020 by Canadian Joint Operations Command (CJOC). The second is the *Canadian Forces Health Services Publication 1 (CFHSP 1), Health Services Planning*, a subordinate publication to CFJP 4-10. Within the U.S. DoD, planning related to medical care eligibility and criteria for admissions are emphasized throughout several DoD and Joint Services publication, as well as service-specific manuals and handbooks. The bulk of specific guidance on the topic of MROE within American military policy, however, is found in U.S. Army-specific doctrine, specifically, *Field Manual (FM)*

[1]As identified in International Humanitarian Law, under Rule 110, "the wounded, sick and shipwrecked must receive, to the fullest extent practicable and with the least possible delay, the medical care and attention required by their condition. No distinction may be made among them founded on any grounds other than medical ones." Customary International Humanitarian Law, Volume II, Chapter 34, Section B, Rule 110, Treatment of the Wounded, Sick and Shipwrecked. (Henckaerts and Doswald-Beck 2005).

4-02, Army Health System (November 2020) and *Army Regulation 40-400, Medical Services – Patient Administration* (July 2014). As a result, we draw heavily upon these two publications in our review of U.S. DoD-specific MROE policy.

4.2 An Overview of NATO Doctrine

To date, NATO has developed over 1300 Standardization Agreements (STANAGs) which define processes, standard military operating procedures, standardized terminology (a NATO dictionary), training standards, and equipment specifications between the member countries of the alliance. NATO Working Groups, with membership drawn from the various NATO members, review and update the STANAGs on a regular basis.

The purpose of a STANAG is to promote interoperability through the provision of common operational and administrative procedures and logistics which aim to permit one member nation's military to utilize the stores and support of another member's military in domains as diverse as communications equipment, information technology, ammunition specifications, etc. Thus, the overarching aim of a STANAG is to promote technical interoperability among member states, which is essential for NATO operations. NATO Standards—a product of the STANAG—are developed by the relevant NATO Working Group, and then reviewed and ratified by each NATO member state, which subsequently will implement the Standard within their respective military force.

4.2.1 An Overview of NATO Medical Doctrine

In the domain of medical support to operations, there are several STANAGs and associated publications which identify minimum standards for a broad range of medically relevant topics including, but not limited to, medical care and capability, aeromedical evacuation and hospital standards, medical staff training standards, and vaccination standards. The following two NATO Standards provide guidance to member nations concerning medical support on operations, including the medical rules of eligibility: NATO Standard Allied Joint Publication 4.10 (AJP-4.10): *Allied Joint Doctrine for Medical Support* (September 2019), and its subordinate publication, NATO Standard Allied Joint Medical Publication 1 (AJMedP-1): *Allied Joint Medical Planning Doctrine* (September 2018).

The stated aim of NATO Standard Allied Joint Publication (AJP-4.10): *Allied Joint Doctrine for Medical Support* is to provide an "overview of the fundamentals required for planning and providing joint medical support on NATO operations" (AJP-4.10, 2019, xxi). The target audience for AJP-4.10 includes both medical and non-medical NATO staff employed at operational level headquarters or formations (AJP-4.10, xxii). The publication explains joint medical support in the context of the political, strategic, and operational spheres, and identifies accepted NATO medical support terminology, principles, guidelines, and responsibilities (AJP-4.10, xxi). Importantly, in the context of this research, AJP 4.10 also provides an overview of the eligibility related principles and policies surrounding medical support, as well as identifies several of the key factors which merit consideration during the MROE development phase.

NATO Standard Allied Joint Medical Publication (AJMedP-1): *Allied Joint Medical Planning Doctrine* was developed primarily for use by medical staff at the operational level and related subject matter experts and is a subordinate publication to AJP-4.10. The aim of AJMedP-1 is to "identify the fundamental principles required to plan and conduct medical support to NATO operations" (AJMedP-1, 2018, VII). As AJMedP-1 is considered the primary NATO source for all medical planning on NATO operations (AJMedP-1, VII), it offers considerably greater depth and detail on the finer aspects of medical planning and medical support to operations than does AJP-4.10. The topic of medical rules of eligibility is also discussed in AJMedP-1, and additional background and information is provided in the publication which will be of assistance to medical planners and staff tasked with developing MROE.

Although medical support to operations remains a national responsibility among troop contributing nations, the joint nature of operations has increasingly resulted in the sharing of this responsibility among NATO Commanders (AJMedP-1, VII). This shared responsibility has brought with it a need for multinational medical solutions for topics as wide ranging and diverse as determining the appropriate force health protection measures, or identifying best practices for liaison with civil organizations providing health care in the region (AJMedP-1, VII). Overlaid upon these considerations is the increasing public expectation for high quality health care and outcomes in the treatment of casualties (AJMedP-1, VII). While the areas identified above are not exhaustive, AJMedP-1 acknowledges that multinational medical solutions create additional planning challenges (VII). As a result, the unifying doctrine contained within AJMedP-1, which covers all aspects of medical support planning, including MROE development, is a vital planning resource for medical staff.

4.2.2 Developing NATO Medical Rules of Eligibility

The MROE are developed during the operational planning process and approved by the operational level Commander in "close cooperation with the medical staff" (AJP-4.10,1-31; AJMedP-1, 3-17). Accordingly, while the Senior Medical Officer may propose the categories of personnel eligible for care, as well as the spectrum of health care that may be provided, the ultimate approving authority for MROE is the operational Commander, a position that is rarely filled by a physician.

The distinction between the responsibility to advise and the authority to approve is important. In an ideal situation, the operational Commander and the Senior Medical Officer will have a close professional relationship, and the Medical Officer will have direct access to the Commander to provide medical advice and guidance freely and openly.

Although MROE may remain similar from one operation or mission to another, they are 'mission specific'. One reason for this specificity is the MROE-related considerations involving the *population at risk* (PAR). The PAR includes all personnel in the area of operations and will directly influence the size and capacity of the medical facilities established in support of the operation (AJMedP-1, 5-5). Depending upon the mission, the PAR may include all NATO and coalition forces, NATO civilians and contractors, the local population (both military and civilian), refugees and displaced persons, as well as detainees and prisoners of war (AJMedP-1, 3-8). Since not all personnel in the PAR will face the same level of risk, not all personnel within the PAR will be eligible to receive the same spectrum of health care (AJMedP-1, 3-8). Consequently, the PAR will inform not only the eligibility categories, in effect, the identification of which categories of personnel are eligible to receive care at the NATO medical facility, but also the spectrum of health care that each group is eligible to receive, as well as any cost recovery mechanisms (AJMedP-1, 3-17).

AJP-4.10 identifies the different categories of personnel who may present for care at NATO medical treatment facilities (as depicted in Fig. 4.1), and thus may need to be considered within the PAR and, subsequently, when developing the MROE. While the list of personnel categories in AJP- 4.10 is not exhaustive,

- NATO military/service personnel;
- host nation military/service/security personnel;
- NATO diplomatic staffs & other NATO civilian personnel;
- NATO Agency civilian employees;
- NATO employed civilian contractors;
- member national and troop contributing nations diplomatic staffs and government employees;
- employees of international organizations;
- employees of non-governmental organizations;
- locally employed civilians (by NATO);
- local population;
- persons distressed at sea;
- persons deprived of their liberty;
- refugees;
- internally displaced persons;
- members of the press/media;
- foreign company employees;
- tourists/travellers.

(AJP-4.10, 2019, 1-31).

Fig. 4.1 Medical eligibility: NATO personnel categories

it demonstrates the need not only for appropriate planning (including the development of MROE), but also the demands for care which could conceivably be placed upon the medical facility, a demand which may well exceed the capability of the facility from both a resource and a staffing perspective (AJMedP-1, 3-17).

In addition to the assessments and calculations, such as the PAR, made during the operational planning process, NATO doctrine identifies that in alliance operations, harmonized MROE, "determined by ethical, legal (including International Humanitarian Law) and operational aspects are crucial to ensure that the instinctive desire of NATO healthcare providers, particularly in Stabilization Operations, is balanced against operational requirements and the over-arching objectives" of the mission (AJMedP-1, 3-17). Thus, during multinational operations, although many, if not all, of the NATO troop contributing nations will be subject to their own national regulations, NATO operations should include harmonized MROE.

Annex C of AJMedP-1 provides a graphic depiction of a representative example of a MROE decision tree tool used during a multinational NATO operation. Once developed, the decision tree, replicated here at Fig. 4.2, is one of several methods used to communicate the MROE to all medical staff. Typically, on operations, a graphic such as this one would be posted in the admissions area of the NATO Medical Treatment Facility (MTF) as well as in the operational headquarters, thereby ensuring knowledge of the MROE eligibility criteria among not only medical personnel, but key NATO operational decision makers.

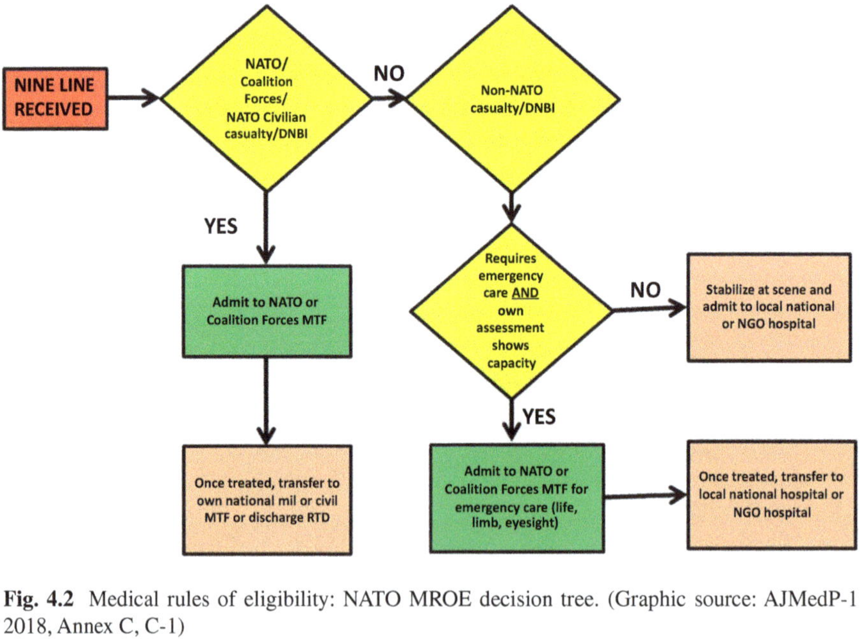

Fig. 4.2 Medical rules of eligibility: NATO MROE decision tree. (Graphic source: AJMedP-1 2018, Annex C, C-1)

In Fig. 4.2, the node at the upper left of the decision tree, titled '*nine line received*' represents a potential incoming casualty to the MTF.[2] In such cases, the initial triage—including the determination of whether the casualty is eligible to receive care, and if so, the level of care—will typically occur at the point of injury, prior to transport to the medical treatment facility. As a result, it may be the on-scene Commander who, in communications with operational headquarters and the MTF, determines whether the casualty is eligible for transport (based upon the MROE).

If the casualty is in one of the three groups that are eligible for full care (indicated in the box immediately adjacent to the "nine line received" box in the decision tree), they will be transported to the MTF. Subsequently, once stabilized, the patient will then be transferred from the NATO MTF to either their own nation's military MTF or into the host nation healthcare system. Alternatively, if the patient is well enough, they may be discharged from the MTF and returned to duty. If the casualty does not meet the medical rules of eligibility as defined in this decision tree, for example, as illustrated in Fig. 4.2, they are either a non-NATO casualty or a Disease Non-Battle Injury (DNBI) case, they may still be eligible to receive medical care. Specifically, if during the assessment by the medical triage officer it is determined that the casualty requires emergency care *and* if there is capacity at the MTF to provide that care, then the casualty *may* be admitted for emergency care. Once stabilized, however, the patient will be transferred to a host nation medical treatment facility.

In the MROE decision tree depicted in Fig. 4.2, only NATO forces, coalition forces, and NATO civilians are *automatically* eligible to receive care at the NATO medical treatment facility. Individuals seeking care who are not members of these eligible groups will typically only receive care at the MTF if the care required is considered emergency care (for example, care that is required to protect life, limb, or eyesight—often simply referred to as 'LLE'), *and* the MTF has the capacity and the capability to provide care for the individual(s).

---

[2] In a combat environment, casualties are often moved from the scene of the accident/injury to the medical treatment facility by helicopter. The '*nine line received*' noted in Fig. 4.2, refers to the standard NATO report used to formally request medical evacuation (medevac) for an injured person (or persons). It is colloquially referred to as the 'nine-liner' because it contains nine lines of information relevant to the medevac request. For example, the radio frequency and call sign of the requesting individual, whether special equipment is required (e.g., a hoist), the level of security at the pick-up site (e.g., possible enemy troops in area, approach with caution), the number of patients (including whether they are ambulatory or require a litter), and the patient nationality and status (e.g., U.S. military, non-U.S. civilian, etc.). Importantly, since the nine-liner is often both requested by and reviewed by non-medical personnel, it does not include any personal medical information.

4.3 A Review of Canadian Armed Forces (CAF) Medical Doctrine

CAF doctrine regarding the provision of medical care during deployed operations is codified in two publications. The first is *Canadian Forces Joint Publication 4-10* (CFJP 4-10), *Health Services*. The CFJP 4-10 is an "interim" publication, released in September 2020 by Canadian Joint Operations Command (CJOC). The aim of CFJP 4-10, the foundational publication on operational Health Services doctrine, is to "set forth doctrine governing the planning and execution of Health Services on operations at home or abroad, within the framework of either a joint, multinational, or multinational joint task force" (CFJP 4-10 2020, iii). Due to the joint and multinational focus of CFJP 4-10, the publication incorporates NATO Health Services doctrine as well as standards published by the American, British, Canadian, Australian and New Zealand Armies Program (ABCANZ) (CFJP 4-10, iii).

The second publication is the *Canadian Forces Health Services Publication 1 (CFHSP 1), Health Services Planning*. A subordinate publication to CFJP 4-10, the CFHSP 1 was released in July 2018 by the Canadian Forces Health Services Group. While both doctrinal publications provide guidance on the provision of medical care during military operations, the CFHSP 1 offers greater detail, depth and clarity for military medical planners and medical personnel, while the target audience for the CFJP 4-10 is non-medical planners and Commanders (CFJP 4-10, iii).

4.3.1 Health Services Planning

Health Services planning is a critical aspect of the operational planning process. Consequently, the mission-specific 'Operations Order' will contain an annex dedicated to the health services component of the mission. In the CAF, the Health Services (HS) plan is normally Annex K of the Operations Order (CFJP 4-10, 3-11). The HS plan is typically drafted by a health services planner, a role normally filled in the CAF by the non-clinical Health Care Administrator or Health Services Officer occupations. CAF doctrine identifies that in addition to being fully integrated into the operational and logistics plan, the HS plan must "conform to the highest level of professional practice, standards and ethics, agreements, and relevant conventions of humanitarian law" (CFJP 4-10, 3-11).

The HS plan details all aspects of medical support to the operation including calculations of the Population at Risk[3] (the PAR) and casualty estimates, both of

[3]Aligning with NATO doctrine, the Population at Risk (PAR) is a key determinant of the CAF Health Services support required and should include all personnel in the operations area, including "CAF, DND civilians and members of other government departments and agencies, contractors, allied forces, local population (military and civilian), refugees, displaced persons, detainees and prisoners of war" (CFHSP 1, 3–8).

which will inform the development of the MROE. Notably, MROE are *not* required if deployed CAF Health Services personnel are only providing medical care to CAF members. The increased frequency of multinational nature of operations, however, has resulted in the expectation that CAF Health Services will often provide medical care to non-CAF personnel (CFHSP 1, 3-22). As a result, MROE development is increasingly a key component of the HS plan since—dependent upon the type of mission and associated mission requirements—the CAF Health Services Unit may be expected to provide health care to non-CAF and non-Coalition personnel. As with NATO MROE, in the CAF, the MROE identifies population groups eligible to receive health care as well as the scope or level of health care that may be provided.

Once complete, the HS plan is reviewed and approved by the Senior Medical Authority, typically a physician. However, since "the duty of care for all personnel, to include the full spectrum of Health Services issues" is a command responsibility (CFJP 4-10, 1-5), it is the operational Commander who will ultimately approve all aspects of the HS plan, including the MROE. Accordingly, it is the Task Force Commander—and not the Senior Medical Authority—that authorizes the Health Services plan, and by extension, the MROE (CFJP 4-10, 1-6). However, a key role of the Senior Medical Authority, in their capacity as "responsible advisor" to the Commander, is to inform and advise the Commander on all matters pertaining to the health of personnel under his or her jurisdiction (CFJP 4-10, 1-6).

4.3.2 Developing CAF Medical Rules of Eligibility

CAF doctrine identifies that medical ethical principles, legal principles, and critically, operational requirements are some of the key considerations that guide the development of the MROE (CFJP 4-10, 2-3). Importantly, at no time during the mission may the health care provided to non-CAF personnel impact upon the ability of the Health Services Unit to perform its primary function—the provision of medical care to CAF members. As with NATO MROE, the CAF MROE are mission-specific, and CAF doctrine explicitly identifies that CAF MROE may differ from the theatre-wide coalition MROE (CFHSP 1,3-22; CFJP 4-10, 2-3).

As was the case in the NATO doctrine, which identified the numerous different categories of personnel that may present for care at NATO medical treatment facilities (see Fig. 4.1), CFHSP 1 identifies that CAF MROE must "clearly state the eligibility for treatment" for each of the identified categories of personnel (CFHSP 1, 3-22). Figure 4.3 presents the personnel categories depicted in CFHSP 1. The CAF's list of personnel categories that may present for medical care is significantly shorter than that presented in the NATO doctrine and depicted in Fig. 4.1 (CFHSP 1, 322; 23). In addition, the CAF doctrine provides considerably more guidance to medical planners with respect to the level of medical services that each of the categories of personnel may be eligible to receive than does the NATO doctrine. CFHSP 1 offers explanatory notes and caveats for each of the personnel categories depicted in Fig. 4.3. Prior to doing so, however, CFHSP 1 clearly identifies that in the deployed

a. DND civilians, other [Canadian] government departments or agencies, and contractors
b. Allied military personnel
c. Third country nationals supporting the operation such as manual laborers, contractors, and private military and security companies who may be entitled under agreements to health care
d. Host nation civilians
e. Host nation security forces
f. Local civilian hires
g. GBA+ identified personnel

(CFHSP 1, 3-22;23)

Fig. 4.3 Medical eligibility: CAF personnel categories

environment "there must be no implied commitment to provide continuous long-term care nor any form of medical treatment beyond the capabilities of existing resources" (CFHSP 1, 3-22).

As with the NATO medical eligibility rules, host nation civilians are not automatically eligible to receive care in CAF medical treatment facilities. However, host nation civilians injured because of the conflict are entitled to receive emergency medical care upon request but are to be transferred into their host nation healthcare system as soon as their condition permits (CFHSP 1, 3-23). The CFHSP 1 notes that the host nation treatment standards, as well as whether there will be a clinical requirement for follow-up or longer-term treatment, should be considered when initiating treatment of host nation civilians at a CAF medical facility (CFHSP 1, 3-23). In essence, if the host nation healthcare system does not possess the necessary clinical infrastructure to either continue or complete the treatment initiated by CAF Health Services, an evaluation of whether it is appropriate for the CAF to commence the initial treatment should be conducted (CFHSP 1, 3-23). Similarly, host nation security forces and local civilian hires are also eligible to be provided with emergency health care at CAF medical facilities. For both these categories of personnel, once their condition has stabilized to a sufficient degree to permit transfer, they too are to be transferred to a host nation health care facility (CFHSP 1, 3-23).

CAF doctrine deviates from NATO doctrine with its inclusion of GBA+ identified personnel as a unique personnel category. Since 1995, the Government of Canada has examined policies and programs through a gender lens (Government of Canada 2016). In 2016, the requirement to conduct a gender-based analysis plus (GBA+) assessment on *all* policies, programs, initiatives, and services was fully implemented across all federal government departments and agencies (Government of Canada 2016). As a result, since 2016, the CAF has applied a GBA+ lens to all doctrine, policy, programs, initiatives, and services to not only ensure equity for women, men, and gender-diverse people, but to ensure that diverse perspectives are included.

The inclusion of GBA+ identified personnel as a separate eligibility category in the CAF eligibility groups is an example of the application of a GBA+ lens to military doctrine. A review of current NATO medical doctrine suggests that, to date, NATO has not yet fully implemented GBA+ into their policy analysis and development processes. For example, the NATO medical support and planning doctrine reviewed herein do not presently include the requirement for a GBA+ analysis within their various frameworks, to include MROE development. CFHSP 1 includes GBA+ personnel as a medical eligibility category with the caveat that "legal advice is required to determine obligations" to GBA+ identified personnel (CFHSP 1, 3-23). However, since members of this population group would also fall into one of the other identified eligibility categories (e.g., host nation civilians, host nation security forces, etc.), it is not immediately clear why CFHSP 1 identifies GBA+ personnel as a separate eligibility category.

In addition to identifying the categories of personnel eligible to receive health care, the MROE must also identify the relative degree and type of medical care that each group is entitled to, whether the health care provided will be subjected to cost recovery, as well as whether each group is entitled to tactical or strategic MEDEVAC.[4] Typically, in CAF operations this is depicted in either an MROE flow chart, similar to the NATO decision tree depicted in Fig. 4.1, or in an MROE Matrix, a representative example of which is provided in Fig. 4.4. The MROE Matrix in Fig. 4.4 identifies each of the personnel categories eligible for care at the CAF medical treatment facility at the top of the matrix, while the left axis of the matrix depicts eligibility for evacuation to a medical facility, as well as the level of health care that each of the categories of personnel are eligible to receive at the CAF medical treatment facility.

Since Canada's military medical health system strives to provide CAF members with a comparable standard of healthcare—whether in garrison or deployed—to that available in Canada, the services and resources available at the deployed Health Services Unit may, in some instances, be more advanced than those available in the host nation. When present, this care dichotomy may raise very real medical ethical dilemmas for military health care providers treating and subsequently transferring local civilian and military personnel into their host nation health care system. These dilemmas are particularly present in areas of intense operations and unpredictable casualty flow, with their resultant pressure on the medical staff to ensure that beds remain available for new admissions. The CFHSP 1 clearly states that patients should never be held longer than clinically necessary and advises medical Commanders and planners that they should "always be thinking about how to empty-out a Medical Treatment Facility" (to ensure bed space) (CFHSP 1, 3-18).

Finally, although the provision of healthcare is the responsibility of the Health Services Unit, as previously noted, the duty of care for all personnel, which

[4] Tactical medical evacuation (MEDEVAC) refers to casualty evacuation and movement *within* the theatre of operations. Strategic MEDEVAC refers to casualty evacuation and movement *from within* the theatre of operations *to outside* the theatre, or from outside the theatre of operations to a third location.

	CAF / DND civilians	NATO	Contractors	Host Nation Security Forces	Host Nation Civilians
Forward / Tactical Air Evac	Yes	Yes	Yes	Yes	If injured by coalition or enemy
Strat Air Evac	Yes	Yes	No	No	No
Primary Care	Yes	Yes	No - Civilian clinic (*if present*)	No	No
Surgical Care	Yes	Yes	If LLE or BI*	If LLE or BI	If injured by coalition or enemy
Specialist Care (e.g., Dermatology, Mental Health, etc.)	Yes	Yes	No	No	No
Dental	Yes	Yes	Emergency care only	No	No
"Allied" Healthcare (e.g., physiotherapy)	Yes	Yes	No	No	No

*LLE = Life, Limb, Eyesight; BI = Battle Injury

Fig. 4.4 Representative example of a CAF MROE Matrix. **LLE* Life, Limb, Eyesight; *BI* Battle Injury

encompasses "the full spectrum of Health Services issues", rests with the operational Commander (CFJP 4-10, 1-5). Additionally, the "operational Commander has the authority to limit the availability of military medical support to third parties" (CFJP 4-10, 1-4). Ideally, this decision is taken in concert with advice from the Senior Medical Authority; however, the final decision regarding the allocation of limited medical resources lies with the operational chain of command. As a result, while emergency treatment of life-threatening conditions is not *normally* denied, the decision as to whether the Health Services Unit is permitted to provide medical care to non-CAF and non-Coalition members is a decision that is made by operational command, and not by senior medical leadership.

4.4 A Review of the United States Department of Defense (DoD) Doctrine

Although the importance of developing eligibility criteria for medical care is emphasized throughout several United States Department of Defense (DoD) and Joint Services publications, as well as several service-specific manuals and handbooks, our review found no overarching pan-DoD guidance on developing medical rules of eligibility. Certainly, while the subject of MROE is mentioned in multiple U.S. DoD Field Manuals, regulations, and publications, in most cases, the MROE were either referenced in passing, or noted in appendices as an important component of medical planning. In general, our team found that while the medical planning process is explicitly prioritized, and the Geneva Conventions are repeatedly cited, few specifics—beyond the importance of compliance with the Geneva Conventions—were provided in most of the U.S. DoD doctrine reviewed. Instead, the doctrine reviewed usually focused on the proper use of and protection of medical supplies, medical personnel, and medical facilities. Explicit guidance on the MROE development process was generally not covered, with U.S. Army doctrine being the exception to this finding. The following section provides a brief review of these documents.

4.4.1 Joint Services Doctrine

While there is no specific joint-services doctrine related to MROE, the most extensive treatment of this subject can be found in policy related to Humanitarian Assistance missions. As an example, *Joint Publication 3-29, Foreign Humanitarian Assistance* (2019) (JP 3-29), explicitly establishes the requirement to provide medical care to civilians. In this doctrine, foreign humanitarian assistance (FHA) is defined as: "DOD activities conducted outside the United States and its territories to directly relieve or reduce human suffering, disease, hunger, or privation" (JP 3-29, 2019, viii). Since these operations involve medical care beyond service-members, NATO members, and allies, there is more explicit consideration of eligibility. Medical forces typically have four distinct missions as part of FHA operations: (1) prevention and/or control of disease outbreaks, (2) evacuation or temporary provision of medical care for the sick, wounded and injured, (3) distribution of supplies and equipment, and (4) assistance in reestablishing host nation health resources and institutions (JP 3-29, E-1). This doctrine emphasizes that medical planning should be integrated into overall response at early stages prior to deployment and that this planning should include the establishment of eligibility criteria for medical care and/or evacuation (JP 3-29, E-4).

Importantly, due to the American health insurance system, there is separate and extensive doctrine related to eligibility in general. DoD Instruction (DoDI) 6025.23 *Health Care Eligibility Under the Secretarial Designee (SECDES) Program and Related Special Authorities* "applies to the Office of the Secretary of Defense (OSD), the Military Departments, the Office of the Chairman of the Joint Chiefs of Staff and the Joint Staff, the Combatant Commands, the Office of the Inspector General of the Department of Defense, the Defense Agencies, the DoD Field Activities, and all other organizational entities in the DoD" (DoDI 6025.23, 2011, 1). This Instruction outline eligibility criteria for the American military health insurance (TRICARE), as well as which care services at state-side MTFs uninsured populations (those not enrolled in TRICARE) may be eligible to receive.

Although DoDI 6025.23 focuses on care delivered within the U.S. and explicitly states that it does not apply to "health care services provided to coalition forces in operational settings, or to allied forces in overseas training exercises and similar activities" (DoDI 6025.23, 1), it does provide a list of population categories that is illustrative for our analysis. The DoDI points to five populations who *may* be eligible to receive care: (1) Foreign military personnel and their dependents; (2) Foreign diplomatic and other senior foreign officials; (3) Other foreign nationals; (4) U.S. Nationals Overseas; (5) U.S. government civilian employees and contractor personnel. The DoDI states that foreign military personnel and their dependents are eligible for *space-available* care, while all other groups are only eligible when there is both space-available *and* compelling or extraordinary circumstances (DoDI 6025.23). As with much of the policy regarding eligibility, cost-recovery and reimbursement are a particular focus given the American military medical system's positioning within the broader American health care market. Still, this Instruction may inform how Commanders develop rules of eligibility for MTFs outside of the U.S.

There are several other DoD/Joint Service policies that mention this type of eligibility criteria, although in general, while planning is explicitly prioritized, no specifics are given. Those policies and documents are listed below:

- Joint Publication 4-02: Joint Health Services
- Joint Publication 3-07: Stability
- DoD Instruction (DoDI) 6025.27: Medical Ethics in the Military Health System
- Defense Health Board: Ethical Guidelines and Practices for U.S. Military Medical Professionals

The two Joint publications identified above (4-24: Joint Health Services and 3-07: Stability) include the development of medical rules of eligibility as part of planning checklists for those in command but do not offer guidance on how the rules should be developed, or what populations should be considered. The final two documents identified above (DoDI 6025.27 and the Defense Health Board's ethical guidelines and practices) focus on medical ethics and provide principles intended to broadly shape MROE according to NATO and Army doctrine (discussed in next section), however, no specific guidance is provided. Additionally, while the Defense Health Board publication references MROE in a case study as an example of a recurring medical ethical issue, it does not offer guidance.

4.4.2 A Review of U.S. Army Medical Doctrine

Much of the specific guidance that was discovered regarding the development of MROE was in U.S. Army-specific doctrine, and in particular, *Field Manual (FM) 4-02, Army Health System* (November 2020) and *Army Regulation 40-400, Medical Services – Patient Administration* (July 2014). Although both publications outline medical eligibility, the information provided within each offers less detail than the CAF doctrine reviewed earlier. Additionally, the term "medical rules of eligibility" is never explicitly used in either publication—this is worth noting since the term is utilized in both NATO and CAF policies. That notwithstanding, both of these Army publications establish who is eligible for care in U.S. Army medical treatment facilities operating outside of the United States.

FM 4-02: Army Health System (FM 4-02) provides the most guidance related to MROE, with specific language[5] and sample matrices. This doctrine explicitly highlights the importance of eligibility of care determinations, stating:

> During unified land operations, one of the most pressing questions is who is eligible for care in a United States Army-established MTF [medical treatment facility] and the extent of care authorized. Numerous categories of personnel seek care in U.S. facilities that are located in austere areas where the host-nation civilian medical infrastructure is not sufficient to provide adequate care (FM 4-02, 2020, 1–13).

Although FM 4-02 identifies the clear need for this type of planning, it does not offer specifics regarding which populations should be considered when making decisions related to establishing eligibility criteria. Moreover, while sample eligibility matrices are provided in FM 4-02, they do not focus on the typical MROE decisions covered in NATO and CAF doctrine. Instead, as depicted in Fig. 4.5, the sample matrices identify specific patient categories and name, as well as their respective eligibility for services ranging from evacuation to dental services. Figure 4.5 provides a three-page sample Eligibility for Medical and Dental Care Support Matrix for treatment in a United States Army Medical Treatment Facility (FM 4-02, 2020, Table 1–2, pages 1–15 through 1–17). As is evident from page 2 of Fig. 4.5, the contractor and employee population feature prominently in the eligibility matrix.

4.4.3 U.S. Doctrine – Conclusion

As was the case with both NATO and CAF doctrine, the U.S. Army doctrine reviewed identified that decisions surrounding MROE development must be informed by national law, international law, the Geneva Conventions, as well as operational and mission requirements. Importantly, and differentiating U.S. Army doctrine from both NATO and CAF doctrine, the MROE matrix is to be both developed and authorized by Command, in conjunction with military lawyers—neither medical advisors nor medical planners are mentioned. According to this policy, eligibility of care criteria should be developed based on the appropriate U.S. and international law and decided "at the highest level possible and in consultation with the supporting staff judge advocate" (FM 4-02, 2020, 1-13). The policy further recommends Department of State and other military staff sections (such as the assistant chief of staff, for example) be included in the determination processes (FM 4-02).

As with the doctrine of both NATO and Canada, U.S. Army MROE are mission specific, with unique eligibility criteria developed for each operation. According to the U.S. DoD doctrine reviewed, other factors impacting the determination of

[5] Still, the term "medical rules of eligibility" (or MROE) is never used.

ELIGIBILITY FOR MEDICAL AND DENTAL CARE SUPPORT MATRIX (DATE) (THIS DOCUMENT IS SUBJECT TO FURTHER VERIFICATION AND AND/OR MODIFICATION)		
Category	***Medical/dental***	***Information/authority****
Multinational military personnel	Yes[1]	The following nations have acquisition and cross-servicing agreements and multinational force compatibility agreements with the U.S. which are administered by *(combatant command): List nations.*
Department of Defense Civilian employees	Yes	Invitational travel order.
U.S. Government employees (non-Department of Defense)	Yes[2]	Invitational travel order.
U.S. Embassy personnel	Yes	U.S. citizens on official business.
U.S. Congressional personnel	Yes	U.S. citizens on official business.
Army and Air Force Exchange Service U.S. citizen employees	Yes	Invitational travel order.
Army and Air Force Exchange Service Local national employees	Yes[3]	U.S. law.
Nonappropriated fund instrumentality morale, welfare, and recreation U.S. employees	Yes	Invitational travel orders.
Contracted college instructors	Yes	Invitational travel orders.
United Nations personnel (includes all personnel employed by the United Nations and its agencies, such as the United Nations High Commissioner for Refugees)	Yes[3]	U.S. law.
Contractor #2 all employees POC: Mr. Michaels (XXX) XXX-XXXX ADMIN: Mr. Johns DSN XXX-XXXX	Yes[3] No[5]	Contractor did not contract for the provision of medical care by military medical treatment facilities. Contractor stated in writing that they contracted with the host-nation medical infrastructure for the required care. NOTE: A separate determination may be required for individual cases, as the individual may be eligible for care under a different provision. Contact Mr. Patrick, DSN XXX-XXXX if additional information is required.
Contractor #4 Mr. Edward Dean (company name classified) POC: Ms. Emory (XXX) XXX-XXXX ADMIN: Mr. Johns DSN XXX-XXXX	Yes	Per Mr. Patrick, Mr. Dean is entitled to full medical and dental support without reimbursement. The terms of the contract and the name of the contracting company are classified. Contact Mr. Patrick, DSN XXX-XXXX, if additional information is required.
Contractor #5 Mr. Michael James (company name classified) POC: Ms. Emory (XXX) XXX-XXXX ADMIN: Mr. Johns DSN XXX-XXXX	Yes[6]	Per Mr. Patrick, Mr. James is entitled to full medical and dental support; however, this care is reimbursable. The terms of the contract and the name of the contracting company are classified. Contact Mr. Patrick, DSN XXX-XXXX, if additional information is required.
Dependents of U.S. active duty or retired military personnel	Yes[4]	Only if space is available and appropriate medical services/care are available in the operational setting. AR 40-400. Contact Mr. Patrick, DSN XXX-XXXX, if additional information is required.

Fig. 4.5 U.S. Army sample eligibility for medical and dental care matrix. (Source: FM 4-02, Table 1–2, Pgs 1–15 through 1–17)

Personnel in custody of U.S. military forces	Yes	Enemy prisoner of war and detained personnel. Extent of care rendered is the same as that provided to U.S. military forces within the geographical area. (Army Techniques Publication 4-02.46, and Field Manual 27-10).
Individual injured as a rssult of military operations	Yes	U.S. and international law (Field Manual 6-27) and status of forces agreements. If the U.S. military injures an individual (such as in an automobile accident involving a military vehicle), the U.S. is responsible for providing immediate care (or paying for local care). Coordinate with Mr. Patrick, DSN XXX-XXXX and Lieutenant Colonel Hall, supporting staff judge advocate, DSN XXX-XXXX.

LEGEND:

* Illustrative in nature only.

1 Multinational force member nations are provided food, water, fuel, and medical treatment pursuant to reciprocal agreements. The amount of food, water, fuel, and medical care provided must be accounted for by the providing nation to the assistant chief of staff, Civil Affairs operations (G-9). Logistical support is not permitted for those nations with whom the U.S. does not have both an acquisition and cross-servicing agreement and multinational force compatibility agreement. However, the acquisition and cross-servicing agreement and multinational force compatibility agreement requirements may be waived for those nations whom the commander, in conjunction with the supporting staff judge advocate, feels are supporting the missions of the command.

2 if not working for, contracted to, or on DOD multinational force compatibility agreement for logistical support, non-DOD U.S. Government employees must pay for meals received at DOD dining facilities.

3 Emergency medical and dental care only. Emergency care is that care required to save life, limb, or eyesight.

4 Space available.

5 Routine.

6 Reimbursable.

Admin	administration
AR	Army regulation
DOD	Department of Defense
DSN	Defense Switched Network
POC	point of contact
U.S.	United States

Fig. 4.5 (continued)

eligibility include command guidance, practical humanitarian and medical ethics considerations, availability of medical assets (in relationship to the threat faced by the force), and the potential training opportunities for medical forces. Overall, the obligations of the MTF are typically framed as stabilization and evacuation, and not the provision of long-term treatment or care. Moreover, although references are made in the doctrine to the (often-cited) eligibility criteria of risk to "life, limb, and eyesight" for civilians in medical crisis, these references mirror state-side doctrine limiting civilian care to emergency care when there is space available at the MTF (FM 4-02, 1-15; 1-17).

Finally, while the U.S. Army's list of personnel categories is neither as extensive nor as explicit as those within either NATO or the CAF, there is an explicit obligation to stabilize *any* casualty—regardless of personnel category—in an emergency, and specifically when there is *risk to life, limb, or eyesight*. The U.S. Army doctrine notes that there may be cases where eligibility is uncertain, for example, in cases where an individual(s) who is not represented in one of the population groups depicted within the MROE arrives at the emergency section of the medical treatment facility seeking care. In such cases, the doctrine notes that the individual(s) must first be medically stabilized, only once this stabilization is achieved will the patient's eligibility for continued care be determined. As is made clear in both NATO and CAF doctrine, the obligation to stabilize the patient does not extend to an obligation to provide continued care.

4.5 Key Similarities Between NATO, CAF, and US DoD Medical Doctrine

Regardless of whether the MROE-specific policies reviewed here fall under the auspices of NATO, CAF, or the U.S. DoD, each policy permits the provision of emergency medical care to individuals within the theatre of operations experiencing an immediate risk to life, limb or eyesight (LLE). However, the level of emergency medical care provided and the location where this emergency medical care is provided may differ, dependent upon both the patient category (e.g., NATO or Coalition Forces, contractor, local civilian) and the current capacity at the military MTF (e.g., the available bed space).

Accordingly, what this means in reality is that the level of medical care provided to a local civilian requiring emergency medical attention as a result of coalition or enemy action may simply be stabilization on-site followed by transfer to their host nation medical facility. Conversely, if capacity permits (i.e., bed space is available at the military MTF), this same injured local civilian may be transferred to the military MTF for emergency treatment prior to transfer to a host nation medical facility. The assessment of whether an injury poses an immediate risk to life, limb, or eyesight—and consequently, whether the individual is eligible for medical care—is not a simple one to make. This is particularly the case when the assessment must be made rapidly, in a potentially hostile and dangerous environment, in many instances by an on-scene Commander who may have little to no formal medical training. It is for this reason that MROE decision trees and matrices that can facilitate rapid decision-making are commonly utilized in military operations.

Decision trees and matrices are imperfect tools for such consequential assessments, however, and thus —and not surprisingly—medical ethical challenges can and do arise when operating in such an environment. Although medical ethics fall outside the scope and aim of this comparative analysis, Michael Gross,[6] George Annas, Janet Kelly, Martin Bricknell and others have written extensively about the medical ethical challenges faced by military medical personnel operating in war and conflict zones and are recommended reading for those seeking more information on this important topic.

[6] See for example, Gross 2021, 2022.

Thus, within NATO, the CAF, and the U.S. DoD the MROE doctrine and policy are, broadly speaking, very similar; any variances between them are nuanced. One area where there is a subtle difference within the policies lies in the relationship between the Senior Medical Authority and the Commander as articulated in CAF and NATO policy, and in particular, the duty of the Commander to consult the Senior Medical Authority on matters that are medical in nature.

CAF doctrine states that the Senior Medical Authority "must have unimpeded access to the Commander to ensure timely intervention in all medical matters that require the Commander's attention, decision or action" (CFJP-4, 1-6). The phrasing of this sentence places the 'responsibility to consult' regarding medical issues upon the Medical Officer, and not upon the Commander. While the argument could be made that the Commander has a parallel responsibility to make themselves available for consultation, the requirement for this "unimpeded access" (of the Medical Officer to the Commander) is not identified within the *Responsibilities of the Commander* delineated in CFJP 4-10 (CFJP 4-10, 1-5). Moreover, it is also not identified within the responsibilities of the *Joint Task Force Commander*, one of whose responsibilities is the responsibility to "determine the medical rules of eligibility" (CFJP 4-10, 1-6). Instead, the responsibility to consult the Commander lies solely with the Senior Medical Advisor. Thus, while implied in CAF doctrine, absent in the published Commander's responsibilities is the *obligation* to consult the senior Medical Officer on medical matters, including MROE (CFJP 4-10, 1-5;6). Accordingly, within the CAF there is a singular upwards flow of responsibility and accountability, from the Medical Officer to the Commander, with no similar corresponding requirement for responsibility or accountability downwards (i.e., from the Commander to the Medical Officer).

The responsibility for non-medical operational Commanders to ensure the health of their force and the authority to apportion limited health resources is not unique to the CAF. The MROE guidance within *NATO Standard AJMed P-1, Allied Joint Medical Planning Doctrine*, identifies that "MROE are approved by the Commander based on advice from the Medical Advisor" (AJMed P-1, 3-18). Thus, in contrast to CAF doctrine, NATO doctrine unequivocally links the Task Force Commander to the medical authority, and clearly identifies the Commander's obligation to seek medical advice on all decisions with a health nexus.

In conclusion, when comparing the medical doctrine within NATO, the CAF and the U.S. DoD several key similarities can be observed. First and foremost, in all three, medical personnel at the medical treatment facility are obligated to stabilize all emergent cases whose condition poses a threat to life, limb, or eyesight. In all cases, however, this is solely an obligation to stabilize; there is no obligation to continue care after stabilization. Furthermore, all three rely upon personnel categories—for example, Allied Forces, Host Nation military, Host Nation civilian—to determine not only the patient's eligibility to receive medical care at the facility, but also the spectrum of care that the patient is eligible to receive. In addition, each doctrine draws upon international humanitarian law, including the Geneva Conventions, when developing MROE. Finally, all three assign the responsibility for developing and approving the MROE to the operational Commander, typically with the involvement of the Senior Medical Authority.

4.5.1 Key Differences Between NATO, CAF, and U.S. DoD Medical Doctrine

Although not extensively discussed herein, it was noted during our review of the doctrine that the U.S. DoD medical doctrine places a greater emphasis on medical insurance and reimbursement than does either NATO or CAF medical doctrine. This is possibly due to the significantly larger population of contractors employed within the U.S. DoD. In addition, neither the NATO doctrine nor the U.S. DoD medical doctrine include 'GBA+' personnel in their patient eligibility categories. Thus, among the three, CAF doctrine is presently the sole doctrine to include 'GBA+' personnel as a separate personnel category in the MROE categories.

4.6 Conclusion

The aim of this chapter was to provide an overview of how medical eligibility decisions are formalized into policy across military contexts. Ultimately, our review of the NATO, CAF and U.S. DoD policy and guidance surrounding medical rules of eligibility on deployed operations revealed more doctrinal similarities than differences. Although this review of medical rules of eligibility was conducted primarily as a comparative analysis (as opposed to an ethical analysis), in this final section, we take the opportunity to briefly highlight some key issues with an ethical nexus.

Given the significant potential for medical ethical challenges to arise when decisions concerning whether—and if so, to what degree—to provide medical care to the ill or injured in theatres of operation are made, we suggest that the lack of attention given to ethics, and in particular medical ethics, in MROE policy is concerning. In the Canadian CFHSP 1, the key CAF doctrine for medical planners, the word ethics appears exactly twice. In the first instance, it is to note that health plans must conform to "professional practice, standards and ethics, agreements and relevant conventions" (CFHSP 1, 3-10). The second instance is found in Annex E to CFHSP 1, an annex which provides Health Services specific guidance on the Law of Armed Conflict. In this annex, medical personnel are reminded that the health care provided to prisoners of war and detainees must be "consistent with medical/dental ethics, applicable law…" (CFHSP 1, E-3). Similarly, NATO's AJMedP-1 mentions ethics twice, in both instances in similar contexts to the CAF citations above, while American policy references medical ethics only once. We suggest that greater attention to medical ethics within doctrine and the greater involvement of ethicists in developing MROE are vital towards ensuring that medical eligibility criteria are ethically sound.

Additionally, given that the medical doctrine for all three groups recommends rapid patient transfer once stabilized (particularly for those patients within host nation personnel categories), we suggest that the decision to transfer, and thus, terminate patient care by military healthcare providers, will present greater ethical dilemmas for medical personnel than will the provision of emergent care. Both CAF and U.S. Army doctrine clearly indicates that patients should never be held at the medical treatment facility for longer than clinically necessary. The pressure on healthcare providers to empty beds is thus a key aspect of the medical doctrine reviewed. This pressure can lead to ethical dilemmas for medical staff. These dilemmas may be amplified—and thus particularly morally distressing to military healthcare providers—in instances when patients are transferred to host nation health systems that may not have the same level of capability to care for the patient.

Finally, one of the greatest challenges for this team throughout the doctrinal review process was sourcing overarching U.S. DoD doctrine regarding medical eligibility on operations. As was noted in this chapter, the U.S. Army had the most coherent guidance on MROE and as a result, this is what we drew upon for this review. Consideration should be given to establishing U.S. DoD medical doctrine at the level of the Defense Health Agency (DHA), thereby streamlining eligibility criteria amongst all U.S. services.

References

Canadian Armed Forces. 25 September 2020. *Canadian Forces Joint Publication 4-10 (CFJP 4-10): Health Services (Interim)*. Custodian: Canadian Joint Operations Command.

Canadian Forces Health Services Group. 31 July 2018. *Canadian Forces Health Services Publication 1 (CFHSP 1): Health Services Planning*. B-MD-101-007/FP-001. Custodian: Canadian Forces Health Services Group.

Government of Canada. 2016. *Gender-Based Analysis*. Web. Available at: https://www.canada.ca/en/treasury-board-secretariat/services/treasury-board-submissions/gender-based-analysis-plus.html.

Gross, Michael. 2021. *Military Medical Ethics in Contemporary Armed Conflict – Mobilizing Medicine in Pursuit of Just War*. Oxford University Press.

———. 2022. The Ethical Challenges of Providing Medical Care to Civilians During Armed Conflict. Chapter 9. In *Health Care in Contexts of Risk, Uncertainty, and Hybridity*, ed. D. Messelken and D. Winkler, 131–143. Springer Nature Switzerland AG.

Henckaerts, Jean-Marie, and Louise Doswald-Beck. March 2005. *Customary International Humanitarian Law. Volume 1. Rules*. Chapter 34, 396. Cambridge: International Committee of the Red Cross. Cambridge University Press.

NATO. September 2018. *NATO Standard AJMedP-1: Allied Joint Medical Planning Doctrine*. Edition A Version 1. Allied Joint Medical Publication. NATO Standardization Office.

———. September 2019. *NATO Standard AJP-4.10: Allied Joint Doctrine for Medical Support*. Edition C Version 1 with UK national elements. Allied Joint Publication. NATO Standardization Office.

United States Army. July 2014. *United States Army Regulation 40-400: Medical Services – Patient Administration*. Headquarters, Department of the Army.

———. November 2020. *United States Army Field Manual (FM) 4-02: Army Health System*. Headquarters, Department of the Army.

Joanne Clifford, is a former Bioscience Officer in the Canadian Armed Forces (CAF). During her military career she had the opportunity to serve in positions as diverse as medical intelligence analyst, national-level public health advisor, ethics advisor, and public health team leader. She has deployed on several operations overseas, including Humanitarian Assistance Disaster Relief missions to Turkey and Pakistan following their respective earthquakes, a peacekeeping mission in Bosnia, and two missions in Kandahar, Afghanistan. Joanne holds a PhD in ethics from Carleton University, as well as a Master of Public Health from the University of Waterloo.

Lt. Col. (ret) Paul C. Eagan, is a former military officer and public health physician in the Royal Canadian Medical Service. He served in numerous roles during his military medical career including the head of the communicable disease control section as well as the director of force health protection for the Canadian Armed Forces. He has extensive training and experience in the areas of reproductive health, public health, occupational medicine, and infectious disease. He helped formulate and deliver the Canadian Armed Forces response to the 2014-2015 Ebola epidemic and has contributed to NATO doctrine on the subject. He has extensive clinical experience in resource-scarce medical situations in Canada's North and in military operational settings in Afghanistan and the Middle East. Dr. Eagan is currently a medical analyst in the Directorate of Health Services Operations, Canadian Forces Health Services, as well as an assistant professor in the Faculty of Medicine, Dalhousie University.

Sheena M. Eagan is an Assistant Professor with the Department of Bioethics and Interdisciplinary Studies in the Brody School of Medicine at East Carolina University. Dr. Eagan holds a PhD in the medical humanities from the Institute for the Medical Humanities at the University of Texas Medical Branch, as well as a Master of Public Health from the Uniformed Services University. Sheena is the creator and president of the American Society of Bioethics and Humanities group for Military, Humanitarian and Disaster Medicine. She has published articles in peer-reviewed journals, military-specific journals, and contributed to edited books on a variety of topics in military medicine. She has also given talks and lectures on her work in North America, Asia, Europe, and the Middle East. Her research and teaching have focused on medical ethics and the history of medicine, with a subspecialized focus on military medicine.

Chapter 5
Ethical and Legal Basis for the Standards of Triage Used in the Russian Military Medical Service

Ivan Kholikov

Abstract Legal norms normally derive from ethical rules, which in case of military medical activity, are reflected in the provisions of international humanitarian law, human right, criminal and military law, various manuals, rules, and procedures adopted both on national and international levels. These provisions confirm that health services must possess immunity from attack and reaffirm that health workers should never be punished for following their ethical obligations to provide care, regardless of the identity or affiliation of the patient.

The obligations of the Russian Federation under IHL are reflected in the standards of triage used in the Russian military medical service. In order to comply with both ethical and legal requirements during the triage it is important to use the most objective criteria to assess the severity of the injury on every stage of medical evacuation. Such assessment is a cornerstone of the triage. While many foreign colleagues approach this problem using morphological, functional, etiological criteria or a combination thereof, in Russia it is more common to use descriptive categories where the complexity of the injury is a stable category and complexity of the condition of the injured is a dynamic one.

Keywords Triage · Casualties · Treatment · Severity of injury · Condition of patient

International Humanitarian Law (IHL) is synonymous with jus in bello and is directed on minimization of sufferings in armed conflicts by protecting and assisting all victims of armed conflict to the greatest extent possible. Jus ad bellum on the other hand refers to the conditions under which states may go to war or to the use the armed force in general. It is necessary to emphasize that the goal of the present chapter is not to judge any specific armed conflict or to discuss the «jus ad bellum»

I. Kholikov (✉)
Institute of Legislation and Comparative Law under the Government of the Russian Federation, Moscow, Russia

S. M. Eagan, D. Messelken (eds.), *Resource Scarcity in Austere Environments*, Military and Humanitarian Health Ethics,
https://doi.org/10.1007/978-3-031-29059-6_5

issues but to focus on jus in bello of which the issues of work of medical personnel (especially performance of triage) constitute significant part.

The norms of IHL oblige military medical personnel to provide medical assistance to all victims of armed conflicts without any distinction. Legal norms normally derive from ethical rules, which in case of military medical activity, are often reflected in various manuals, rules and procedures adopted both on national and international levels. Moreover, a moral code of common values known as the "Ethical Principles of Health Care in Times of Armed Conflict and Other Emergencies" was elaborated by the ICRC 6 years ago. This document aims to ensure the safety and protection of health care workers as well as patients in situations of armed disputes, and was endorsed by a number of international organizations, including the ICMM. Later the UN Security Council unanimously adopted Resolution S/RES/2286 on Health Care in Armed Conflict reaffirming principles of international human rights and humanitarian law that provide health services immunity from attack and demanding that states and all parties to armed conflict comply with their provisions. The resolution reaffirms also that health workers should never be punished for following their ethical obligations to provide care, no matter the identity or affiliation of the patient.

The assessment of the trauma severity is one of the oldest problems of medicine. It is known that in ancient Egypt (IV–III century B.C.) the head injuries were divided into light, severe and hopeless. Such assessment by severity is the basis of military medicine in general and of war surgery in particular, because it comprises the main criteria of allocation of patients to a certain group which is the basis for triage. These were the reasons why Napoleon's battlefield surgeon Dominique Larrey suggested to first evacuate from the battlefield those who were severely wounded irrespective of their military ranks, and the famous Russian war surgeon Nikolay Pirogov in the middle of XIX century allocated hopelessly wounded, severely wounded and lightly wounded.

The obligations of the Russian Federation under IHL are reflected in the national manual on IHL for the Armed Force and in standards of triage used in the Russian military medical service. The manual is very definite on the statement that military medical personnel is obliged to provide medical care to all victims of the armed conflict without any distinction. A key component of the delivery of medical care to mass casualties, which are typical for both armed conflict and disaster situations, is the process of triage, from the French word "trier", meaning "to sort". This concept was introduced by Dominique Larrey, and has since become a cornerstone of military medical care (Frykberg 2005). It involves matching the limited resources to the needs of casualties by assigning those who are most seriously injured to receiving priority care. This requires rapid identification of the severely injured in order to apply these resources most appropriately.

Since the manual was approved and entered into force by the Order of the Minister of Defence of the Russian Federation, it formed an integral part of the military legal system and is compulsory for the Russian Armed Forces personnel. Moreover, the manual refers to the adopted ethical rules of medical professionals. And in order to comply with both ethical and legal requirements during the triage, especially in the situation with scarce resources, it is important to use the most objective criteria to assess the severity of the injury on every stage of medical evacuation.

The mass casualty scenarios may vary depending on where the disaster took place (city or rural area), how far is the nearest treatment facility, whether there is a limited access to the injured and their evacuation (airport, stadium, metro), what caused a disaster (explosion, fire, building collapse, flood, traffic accident, etc.), which damage factors prevailed (blast wave, splinters, flame, carbon monoxide or other toxic substances), how many and which healthcare facilities are available for the medical response to the disaster. It is necessary to remember that the number of dead on the disaster site can reach up to 30%.

To large extent the specialties of medical facility's initial reaction to mass inflow of patients is determined by the amount and quality of pre-hospital aid: whether it is provided by medical personnel or not (in case of self or mutual aid), whether the triage algorithms are used on site (START, Triage Sieve, SALT, VPH and other), whether the casualties are transported by any available means of transport or by ambulances and whether the severe casualties are equally distributed among the specialized medical establishments or all casualties are transported to the nearest hospital regardless of the severity of their injuries.

The development of the disaster with lots of casualties has its own logical sequence, understanding of which is critical in the overall efficiency of the response of treatment facilities to the mass income of patients. From the trigging moment of the disaster the chaos phase follows which lasts for 20–30 min – up to the time when rescue services arrive. There is no medical assistance in this phase. The casualties with light (outpatient) injuries and acute stress reaction either by their own or with the help of relatives (or witnesses of the disaster) reach the treatment facilities. By mistake, they can fill the deficient beds of the intensive care, surgical and traumatological units.

The second phase is the organizational one. It starts from the moment when first medical (emergency and/or disaster medicine) team(s) arrive(s). In this phase the principal triage, measures on liquidation of life-threatening consequences of injuries are performed, the approximate number of casualties is specified, and the hospital duty services are informed. Depending on the hospital's remoteness from the disaster's site the medical teams either adhere to the principle «scoop and go» which priorities fast delivery, or to the principle «stay and play», which means the delivery of full anti-shock aid on-site. The duration of this phase depends on the scale of the disaster and rapid reaction of the rescue services. The hospitals have 40–60 min (!) to prepare for the income of most severely injured casualties. The last phase of the disaster – liquidation of the consequences of the disaster – starts after 2–3 h from the initial phase and normally lasts up to 48 h. The hospitals admit the casualties with light and/or acute stress-reaction which abandoned the disaster area without medical assistance.

During last decades significant experience of rendering surgical and traumatological aid in peacetime disasters was accumulated in Russia (Bykov et al. 2009). Also, important ideas of complex response to mass casualty scenarios were formulated by the colleagues from Germany, France, Israel and USA (Boyd et al. 1987; Champion et al. 1983; Cowley et al. 1985; Frykberg 2005; Galvagno Jr et al. 2019; Zeindler et al. 2021).

According to all specialists, the key role in efficient admission of large number of casualties is played by the plan of emergency response to disaster. The algorithm of medical establishment's work must be practiced during preliminary preparation and regular training according to various scenarios (terrorist act, mass traffic accident, fire, flood, poisoning, etc.) In all these scenarios the admission of casualties begins from triage.

While many foreign colleagues approach this problem using morphological, functional, etiological criteria or a combination thereof, in Russia it is more common to use descriptive categories where the severity of the injury is a stable category and severity of the condition of the injured is a dynamic one. To characterize the severity of injury the following descriptive categories are used: light injury, medium severity injury, severe injury and extremely severe injury.

Similarly, both in the USSR and then in Russia the four-stage descriptive gradation of severity of condition is continued to be used: satisfactory condition, medium severe, severe and extremely severe. In rare cases the fifth gradation is added to these four – critical or terminal condition.

The differences in methods of assessment of severity of condition are stipulated by broader sphere of their application: triage of casualties on the site of injury or in hospital admission units, objectivization of condition severity assessment in the course of treatment, diagnostic of poly-organ disfunction or insufficiency, prognosis of complication development, treatment outcomes, various types of monitoring.

The first issue which needs to be addressed on site when rendering pre-hospital aid to casualties is the assessment of severity of injury to make the triage decision. At the beginning of 1990s the Surgeon General of the Russian Federation Armed Forces gave the task to develop the system of objective assessment of the severity of traumas and injuries which could be used first of all in war surgery but also in damage surgery of peacetime. It was named by the acronym "VPH", and the methodology used to create it was based on elimination of disadvantages of previous approaches. This goal was achieved by separate assessment of trauma severity using the two parameters: severity of damages and severity of the condition along with introducing high level of objectiveness and new features such as polycriteriality and universality to the particular assessment methods (Bykov et al. 2009).

One of the main features of the VPH system is its universality, in other words the possibility to objectively assess the severity of not only isolated but also multiple, joined and combined injuries.

Another feature of the system is its polycriterial scales which allow to assess the trauma severity basing not on one criteria of the most likely outcome but on three: the probability of lethal outcome, disability and duration of combat (or work in case of peacetime) capability loss. This allows to get a better understanding of the final outcome of treatment. The comparative efficiency of existing scales is their ability to objectively assess the severity of injuries where the criteria "survived/died" does not work. Let's consider the example of sprain fracture of the lumbar vertebra with damage of the spinal cord. Patients with such trauma do not die but become seriously disabled and spend long time in hospitals. In case of mono-criterial "survived/died" or "shock developed/not developed" approach the severity of such injury will not be significant. However, if the poly-criterial approach is used with probability of lethal outcome criteria being zero, the big input into the index of trauma severity will be given by the criteria of probability of disability and duration of work capability loss, and the injury will be assessed as severe. This approach plays significant role in medical expertise.

When casualties arrive to various stages of qualified medical aid a number of issues and tasks arise which require objectivization especially when rendering aid to the casualties with severe combined wounds and traumas. First of all, it is necessary to objectively define the condition of the casualty, the reasons stipulating it, the life supporting systems of the injured that have been badly affected and the degree of expression of the revealed disorders.

The optimal solution for achieving these goals is the diagnostic algorithm which comprises the dialogue part of the disease record. Its basic principle is the system-by-system examination by active search of the most informative syndromes which reflect the functioning of the main life support systems of the body. This algorithm includes not only clinical symptoms but also symptoms obtained by invasive instrumental, X-ray, ultrasonic and laboratory methods. The important part of diagnostic algorithm is the scale of objective assessment of casualties' condition – VPH-SP. Realization of diagnostic algorithm ends by defining the diagnose which in case of combined injuries and traumas is formulated by particular body parts. Then each injured body part is assessed using the VPH-P(MT) scale for mechanical traumas, the VPH-P(OT) for gunshot wounds, the VPH-P(R) for non-gunshot injuries. Thus, the index of each injury is defined which allows to objectively assess their severity and to identify the main one. Further all indexes are summed up and the general index VPH-P is obtained.

For further detailing of diagnosis, forecast of complications development, determination of most rational tactics of intensive care and surgical treatment additional indexes are used, such as VPH-SU, VPH-SGE(P), VPH-SGE(D) and VPH-HT. As a result, nosological diagnosis is formed by all body parts in the form of a text with the severity of injuries indicated in points using the VPH-P scale, the severity of condition indicated in points using the VPH-SP scale and the total severity of trauma indicated traditionally (light, medium severe, severe and extremely severe) basing on the summary of the above-mentioned points (Bykov et al. 2009).

The VPH system has been used for more than 20 years in the Russian Military Medical Academy and other treatment institutions of the Ministry of Defence in peacetime and in armed conflicts and demonstrated its high efficiency in reaching the four basic goals:

1. Identification of injuries for objective trauma classification, comparative analyses of wounded and injured treatment outcomes, of methods of treatment, results of treatment institutions activity, results of research on trauma problems.

2. Performance of triage of casualties with injuries while rendering emergency pre-hospital aid to them.
3. Assessment of general condition of casualties at various stages of treatment, monitoring their condition in the process of intensive care, objective evaluation of indications for urgent and deferred operations, objectivization of tactical treatment decisions.
4. Forecasting of options of traumatic disease course, likelihood of complications and outcomes of treatment, diagnosis of poly-organ disfunction or insufficiency.

As far as the triage on hospital stage is concerned its main idea is to distribute the casualties according to the severity of their condition and type of injury into the similar groups based on place, urgency of treatment (in-hospital triage), as well as in cases of transfer to other treatment facilities according to means and sequence of further evacuation (evacuation-transport triage). The aim of the triage is to provide the efficient medical care to maximum number of casualties that have chance to survive. The focus of medical care can no longer be on each individual, but must shift to the population as a whole. The standard goal of providing the greatest good for each individual patient must change in a mass casualty setting to the greatest good for the greatest number.

Traditional approach to triage allocates four/five groups of casualties basing on the severity of injury and/or treatment priorities. There are special hospital triage scales such as Australian Triage Scale (ATS), Canadian Triage and Acute Scale (CATS), Triage Score (TS), Manchester Triage System (MTS), American Emergency Severity Index (ESI), A Severity Characterization of Trauma (ASCOT), Rapid Emergency Triage and Treatment System (RETTS), Medical and Emergency Triage and Treatment System (METTS), which are based on the evaluation of 4–5 physiological functions of the patient. Also triage revised trauma score (T-RTS), Vittel criteria, Mechanism/Glasgow Coma Scale/Age/Systolic blood pressure score (MGAP), and the new trauma score (NTS) are used. These scoring schemes allow a rapid and accurate prognostic assessment of the severity of potential lesions (Cassignol et al. 2019).

Most frequently the following parameters are assessed by using known mnemonic schemes A-B-C-D-E or A-B-B-C-S: level of unconsciousness, breathing frequency, pulse, systolic blood pressure. Less frequently the permeability of respiratory tract, capillary refill speed, location of injuries are assessed. Each physiological condition is given certain number of points which are summed up. Depending on the summary the 4–5 triage groups are allocated: red (severe and extremely severe condition with life threatening consequences of injury) – they require urgent admission to operation/antishock room; orange (severe and extremely severe condition with life threatening consequences of injury) – they require urgent admission to ICU; yellow (medium severe condition without life threatening consequences of injury) – they require advance diagnostic and urgent/deferred surgical treatment; green (satisfactory condition) – medical aid can be provided in out-patient mode; black (clinical death or agony condition) – the casualty died or chances to survive very (Frykberg 2002).

The experience from real events shows that it is extremely difficult to correctly allocate the triage groups and define treatment priorities by just rapid overview of casualties on triage site (especially in cases of polytrauma and mine and blast injuries). Two principal mistakes are so-called undertriage or overtriage.

Undertriage is the inappropriate assignment of critically injured victims with life threatening problems to a delayed category. Undertriage is considered a medical problem that could lead to adverse consequences or death from treatment delay.

Overtriage is the assignment of non-critical casualties to immediate care. With small numbers of patients at any given time, it represents an economic, logistical, financial, and administrative strain on hospitals, personnel and resources, but not a medical problem that could threaten lives. However, it is different in a mass casualty scenario.

Only knowledge of the laws of mass income of casualties and the obtained experience allow to reduce the amount of wrong triage decisions. When planning the amount of in-hospital surgical/trauma aid in case of liquidation of medical consequences of disaster it is necessary to keep in mind that among the incoming casualties only 10% will require resuscitation and emergency (lifesaving) operations, 30% will require urgent and postponed surgical treatment in hospitals and 60% will only require out-patient aid.

Thus, for instance, in order to provide surgical/traumatological aid to 30 ± 5 casualties the hospital should have 3–4 unoccupied resuscitation beds, 3–4 operation rooms, 3–4 surgical brigades (one brigade consists of two surgeons, an anesthesiologist, an operation nurse and an anesthetist nurse); 15–20 beds in surgical wards and 2 rooms for rendering aid to out-patient casualties.

In peacetime the triage in the hospital is targeted not precisely on definite allocation of four or five groups of patients but more on separation of lightly injured and just scared people that may block the efforts of medical professionals on saving lives of heavily injured casualties. The mass inflow of patients makes medical personnel meet the unusual types of injuries and unique clinical problems that they normally do not come across in their everyday practice. Training for working in such situations requires not only special planning and constant practice but, more important, different thinking (Frykberg 2002).

The mass inflow of patients can be triggered not only by disasters or combat activities but also by infectious diseases outbreaks. Such scenarios also require triage. Currently all people of our planet are witnessing the absolutely new infection caused by SARS-CoV-2 virus. This infection does not only challenge the human life but also threatens to completely change the domestic, administrative, economic, social foundations of human activity (Chernogor et al. 2020; Kholikov 2020). This happens despite the existence of sophisticated and well-arranged system of prevention of biological threats and the sometimes close to ideal work of sanitary and anti-epidemic services. The best healthcare systems fail to cope with the inflow of severe infectious patients with during COVID-19 crisis which was exacerbated by a shortage of specific medical equipment and resources, including PPE, ventilators, ICU beds, and physicians, nurses, and technicians (Annas 2020).

In such difficult times for mankind, it is hard to underestimate the importance of adequate functioning of clinical part of healthcare system based on rational organization of work and the heroic deed of each medical professional in triage, diagnostic, treatment and care for mass patients.

The main task of the triage of patients with infectious diseases on the pre-hospital stage is to define the uniform groups of patients with the goal to organize rational implementation of optimal treatment-diagnostic schemes and to prevent the in-hospital spread of infection. Thus, the pre-hospital stage triage is aimed to allocate the following groups of infectious patients:

1. based on preliminary diagnosis (group of patients with the major damage of particular organs and systems);
2. based on the severity of condition and transportability;
3. based on evacuation destination;
4. based on epidemiological hazard.

Epidemic outbreaks or epidemics are mostly caused by some single infection (mono-infection). In such situation each patient is given a diagnosis at the early stage of disease which is based on the main clinical syndrome revealed (pneumonia, acute intestinal infection, meningitis and others). If the outbreak is caused by several infections (poly-infection) it is necessary to indicate the combinations of syndromes (for instance, acute intestinal infection combined with hepatitis) (Ogarkov et al. 2005).

The severity of condition is defined basing on clinical-pathogenic indications of the patient and can be extremely severe, severe, medium severe and satisfactory. The risk factors of critical conditions and complications development must also be marked.

Transportability of infectious patients is determined by the severity of condition but also by the conditions of transportation (distance, duration, means of transport, conditions of the road if transported by ground vehicle, flight altitude in case of air transportation and other factors).

In accordance with the severity of condition and risk factors availability the medical aid is rendered and the sequence and destination of evacuation are set (infectious hospitals, infectious and therapeutic units of multi-profile hospitals, etc.)

The triage of patients infected by highly contagious, contagious, low contagious and non-contagious is performed in accordance with the level of epidemiological hazard. Patients with symptoms of respiratory organs are mostly hazardous for other people, that is why strict anti-epidemic measures are required. Patients with the damage of digestive and other system organs are less hazardous from epidemiological standpoints.

Depending on a level of epidemiological hazard the four groups of infectious patients are allocated taking into account the mechanism of transfer of pathogen of each nosological form:

1. patients with highly contagious infections (plague, cholera, small pox, SARS-CoV, MERS, COVID);
2. patients with contagious infections (typhoid fever, paratyphoid, meningococcal disease, dysentery, typhus, anthrax and other);

3. patients with low contagious infections (malaria, brucellosis, tularemia, etc.);
4. patients with non-contagious infections (group of diseases caused by botulin and other bacterial toxins, leptospirosis, Hemorrhagic fever with renal syndrome and similar).

Patients with diseases from three first groups are sent to infectious or other specially re-profiled units of hospitals where they are treated until cured or in case of complications when treatment, as a rule, takes 30–45 days, until they are not hazardous for other people. Patients from the fourth group can be sent to therapeutic hospitals (Zhdanov et al. 2013).

Triage can be performed on the site of medical organization (infectious unit, field infectious hospital, re-profiled institution) during the admission of infectious patients which is successfully conducted when rendering care to COVID patients. The triage of infectious patients is the main component of organization of accommodation, examination and treatment of patients who are already located in the corresponding units and must be performed from the moment of their admission until their discharge from hospital.

The COVID-19 pandemic along with challenge gave people chance to reconsider many aspects of our lives, including those related to biological security. Already at this stage it becomes clear that not only healthcare and infectious prevention systems but also each single physician no matter what is his or her specialty must be ready to work in the conditions similar to combat and possess the competencies of military doctor in general and infectiologist in particular (Zhdanov et al. 2020).

Written documentation of triage decisions and casualty management is an essential tool for maintaining continuity of care during a mass casualty event. The chaotic environment can easily lead to losing track of casualties, and to critical omissions of treatment or redundant triage and management, as casualties move to successive echelons of care. This documentation also allows a post-event analysis of triage decisions, casualty management and casualty outcomes, from which important lessons can be derived to improve performance in future events. The quality and accuracy of triage decisions, and the impact of these decisions on casualty outcome, are key elements of any post-event analysis (Stein and Hirshberg 1999).

The results should be used to further train triage officers, and add to our body of knowledge of triage decision-making. This will allow us to strive toward the most effective triage in mass casualty disasters (Frykberg 2005).

5.1 Conclusion

The principles of triage are not taught in many medical schools or in residency training. This is why education and training assume major importance in the care of mass casualties from any form of disaster, in view of how different the decision-making must be if the salvage of life is to be maximized (Rignault 1992).

Triage errors are especially likely with the nature of injuries seen in major disasters. These injuries tend to be either very different from those normally encountered in daily medical or surgical practice (i.e. acute radiation syndrome, toxic chemical exposure, blast lung injury, unusual infectious agents), or the extent and severity are much greater than typically seen (i.e. traumatic amputations, multiple shrapnel wounds, abdominal evisceration, decapitation). There may be unusual combinations of injuries that make triage decisions especially difficult, such as a ruptured spleen with hemorrhagic shock in a casualty also contaminated with toxic chemicals or with COVID-19. Such multidimensional injuries should be anticipated in disaster planning and drills. Medical personnel, especially triage officers, must be educated as to their recognition and approach to management, to avoid being caught off guard, and to maximize triage accuracy (Frykberg 2005).

The VPH methodic system used in Russian Military-Medical Service is based on the principle of diagnostic algorithm and is a reliable and efficient instrument of assessment of the condition of casualties with severe injuries. It is extremely sensitive in relation to the forecast of the final outcome and complications development probability. This approach allows to formulate the clinical image of a particular casualty in a particular time period and to optimize realization of the principle of individual approach in surgical treatment and intensive care. The application of multi-level and consequent "VPH" system of trauma severity assessment can significantly improve both the triage results and the further treatment results, especially in casualties with severe combined injuries and traumas.

The system of triage used for patients with infectious diseases in the Russian military proved to be efficient for COVID-19 patients as well.

In general, the standards of triage used in the Russian Military Medical Service are fully compatible with the generally adopted ethical principles and have solid legal basis. However, it is necessary to be able to timely adopt them in accordance with global contemporary challenges and threats to public healthcare.

References

Annas, G.J. 2020. Rationing crisis: Bogus standards of care unmasked by COVID-19. *The American Journal of Bioethics* 20 (7): 167–169. https://doi.org/10.1080/15265161.2020.1779858.

Boyd, C.R., M.A. Tolson, and W.S. Copes. 1987. Evaluating trauma care: The TRISS method: Trauma score and the injury severity score. *The Journal of Trauma* 27 (4): 370–378.

Bykov I.Y., E.K. Gumanenko, I.M. Samokhvalov, and others. 2009. War surgery. National manual. – M.: GEOTAR, p.655–690.

Cassignol, A., T. Markarian, J. Cotte, J. Marmin, C. Nguyen, M. Cardinale, V. Pauly, F. Kerbaul, E. Meaudre, and X. Bobbia. 2019. Evaluation and comparison of different prehospital triage scores of trauma patients on in-hospital mortality. *Prehospital Emergency Care* 23 (4): 543–550. https://doi.org/10.1080/10903127.2018.1549627. Epub 2019 Jan 7.

Champion, H.R., W.J. Sacco, and T.K. Hunt. 1983. Trauma severity scoring to predict mortality. *World Journal of Surgery* 7 (1): 4–11.

Chernogor, N.N., A.I. Zemlin, I.V. Kholikov, and I.A. Mamedova. 2020. Impact of the spread of epidemics, pandemics and mass diseases on economic security of transport. *E3S Web of Conferences* 203: 05019. https://doi.org/10.1051/e3sconf/202020305019.

Cowley, R.A., W.J. Sacco, W. Gill, H.R. Champion, W.A. Knaus, E.A. Draper, and D.P. Wagner. 1985. APACHE II: A severity of disease classification system. *Critical Care Medicine* 13 (11): 818–829.

Frykberg, E.R. 2002. Medical management of disasters and mass casualties from terrorist bombings: How can we cope? *The Journal of Trauma* 53: 201–212.

———. 2005. Triage: Principles and practice. *Scandinavian Journal of Surgery.* 94 (4): 272–278. https://doi.org/10.1177/145749690509400405.

Galvagno, S.M., Jr., M. Massey, P. Bouzat, R. Vesselinov, M.J. Levy, M.G. Millin, D.M. Stein, T.M. Scalea, and J.M. Hirshon. 2019. Correlation between the revised trauma score and injury severity score: Implications for prehospital trauma triage. *Prehospital Emergency Care* 23 (2): 263–270. https://doi.org/10.1080/10903127.2018.1489019. Epub 2018 Aug 23.

Kholikov, I. 2020. The spread of epidemics, pandemics, and mass diseases as a contemporary global challenge. *Pathways to Peace and Security* 2 (59): 27–40.

Ogarkov, P.I., V.V. Vasil'ev, A.N. Kovalenko, A.I. Solov'ev, and A.R. Volgin. 2005. The experience of preventing infectious diseases during liquidation of tsunami consequence in Indonesia. *Voenno-Meditsinskii Zhurnal* 326 (6): 4–8.

Rignault, D.P. 1992. Recent progress in surgery for the victims of disaster, terrorism and war. *World Journal of Surgery* 16: 885–887.

Stein, M., and A. Hirshberg. 1999. Medical consequences of terrorism: The conventional weapon threat. *The Surgical Clinics of North America* 79: 1537–1552.

Zeindler, M., F. Amsler, and T. Gross. 2021. Comparative analysis of MGAP, GAP, and RISC2 as predictors of patient outcome and emergency interventional need in emergency room treatment of the injured. *European Journal of Trauma and Emergency Surgery* 47 (6): 2017–2027. https://doi.org/10.1007/s00068-020-01361-w. Epub 2020 Apr 13.

Zhdanov, K.V., S.M. Zakharenko, and A.N. Kovalenko. 2013. Delivery of health care to contagious patients during the pre-hospital stage. *Voenno-Meditsinskii Zhurnal.* 334 (1): 39–43.

Zhdanov, K.V., K.V. Kozlov, K.S. Ivanov, Yu.I. Lyashenko, and V.S. Sukachev. 2020. Military infectiology for biosecurity control. *Kubanskii Nauchnyi Meditsinskii Vestnik* 27 (4): 38–50. https://doi.org/10.25207/1608-6228-2020-27-4-38-50.

Colonel (ret.) Ivan Kholikov is a graduate of Military University, Moscow. He participated in such international campaigns as United Nations Mission in Angola – 1996, Multinational Operation in Kosovo – 2001 and United Nations Mission in Chad – 2009. In 2014 he was in charge of the Russian military Ebola response team in the Guinea Republic. Having completed 24 years in the military (of which 17 in military medical service) he retired in March 2016 from active-duty service being decorated with a number of awards and medals for the distinguished service.

Currently he occupies the positions of Professor of the Chair of International and European Law at the Institute of Legislation and Comparative Law under the Government of the Russian Federation and Professor at the Research Institute of Occupational Health.

Professor Kholikov is a Doctor of Law, author of numerous publications on international and military law, organization of healthcare and international cooperation.

Chapter 6
The Phenomenon of Allocation: Military Pathways in the Light of Biomedical Ethical Principles

Dirk Fischer

Abstract The challenge of the covid-19 pandemic brought to consciousness the severe ethical problem of apportionment of limited resources in medicine. In pre-covid times, this challenge was more the concern of a minority of key specialists, rather than an issue that troubled the majority of medical ethicists. While military medicine was quite familiar with the phenomenon, allocational problems resulting from a limited number of personnel or material resource troubled civil medical care only in exceptional cases (e.g., disaster medicine, transplant medicine). By now, the allocation of scarce medical resource has become a widespread major concern. The tragic choices linked to allocational strategies (e.g., triage) has to be ethically justified. Based on general terminological considerations, the relevance of biomedical ethical principles (autonomy, non-maleficence, beneficence, justice) will be discussed. Though the latter do not seem to matter at first sight in this context, they deeply affect both physicians' and patients' self-understanding as well as the resulting physician-patient relationship. Therefore, allocational ethics should be an integral part of military as well as civil teaching programs in medicine.

Keywords Allocation · Allocational problem · Allocational principles · Allocational ethics · Tragic choice · Triage

6.1 Introduction

The problem of limited resources in medicine might lead to tragic choices. The nature of these tragic choices is to decide who receives a specific medical treatment should two or more patients be competing for it. In reality, where all or many

D. Fischer (✉)
Teaching and Research Unit for Military Medical Ethics, Bundeswehr Medical Academy, Munich, Germany
e-mail: lfwme@bundeswehr.org

S. M. Eagan, D. Messelken (eds.), *Resource Scarcity in Austere Environments*, Military and Humanitarian Health Ethics,
https://doi.org/10.1007/978-3-031-29059-6_6

patients are equally in need of therapeutic measures, the act of apportionment can be characterized as a life or death decision. As resources are limited, so that equal treatment of all patients in need is impossible, the use of material and the deployment of personnel requires specific selection criteria. In these cases, the process of distribution has an allocational character: while distribution in general does so without specific selection criteria, the latter play a decisive role for allocation.

The worldwide urgency that allocation and allocational problems would gain in 2020 and 2021 was one of the most severe and shocking moments of our time. The world we knew before the pandemic was different from the one we are living in today. The fact that allocational ethics became far more prominent in significance demonstrates this clearly.

In pre-covid times, the ethical problem of allocation was more the concern of a minority of key specialists, rather than an issue that troubled the majority of medical ethicists. The areas of military medicine and disaster medicine were the main operational areas where the topic of allocation could be observed. Many severe ethical problems that military medical personnel have to face in contemporary employments owe an allocational character. It is therefore not surprising that military medical ethicists discussed allocational problems regularly over the last decade. In fact, allocation and allocational problems play a crucial role in the sphere of military medical ethics, e.g. when it comes to triage in a mass casualty incident or the treatment of the civil population in a country of operation.

The recent pandemic exposed the scarcity of the medical resources available to deal with such a sudden global crises. This is true even for medically highly developed countries. The allocation of scarce medical resources is now an area of major international concern. It has become widely discussed in society, and the term *triage* has become part of everyday language. Today most people have come across the concept of triage and the moral challenges that it presents. Before the pandemic, the amount of scientific publications on the topic was rather limited. This was true for both the civil and military context.

Due to this development, it seems necessary to highlight some aspects of the current debate, which underline the importance of allocational ethics for both civil and military medical ethics. Of course, this attempt to illuminate the phenomenon cannot meet the claim of solving the problem of allocation in medical ethics.

6.2 Terminological Aspects

The phenomenon of allocation and its ethical challenge are strongly linked to the major philosophical question of distribution and justice. Since Plato and Aristotle, philosophers tried to illuminate this severe problem, and finding a fundamental solution has troubled thinkers of various disciplines until the present day. Until today, the distinction between *iustitia commutativa* and *iustitia distributiva* plays a decisive role. Brought up by Aristotle and taken up by Thomas Aquinas this theoretical approach provides the basis for most theories of justice (Höffe 2008).

While the relation of peers, that is to say persons who interact in a mutual or reciprocal way, characterizes *iustitia commutativa iustitia distributiva* reflects of an unequal relationship. A specific trait of this unequal relationship is that at least three persons or parties are involved. The distributing one and the receiving ones. *Iustitia distributiva* raises the question of how a fairly distribution can be done in this case. For Aristotle the idea of merit or worthiness is the crucial point of view, which forms the basis of a fair proceeding. Yet, Aristotle himself was quite aware of the fact, that merit and worthiness are quite difficult to define and culturally dependent. Criteria like status, neediness, performance or equality lead to quite different answers to the question of a just and fair distribution (Maio 2017, 167–170). It is not surprising that liberalism, egalitarianism, utilitarianism or contractualism came to very different results.

So far, we have sketched the problem presupposing an unlimited availability of the good to be distributed. The question of distribution becomes much more difficult if the number of resources undermine the number of those who compete for them. This is where allocation comes into play. The term distribution (in German *Ver*teilung), refers to a general hand-out without feeling the pressure of a limited resource, whereas the term allocation (in German *Zu*teilung) implicitly reflects the scarcity of resource. The scarcity of resources, which can be of a material, financial, personnel or temporal nature for instance, in distributive theory, gives rise to allocational problems.

Based on the different aspects of the phenomenon we have seen so far, I would like to propose a set of definitions, which allow circumscribing the related problem area:

Allocation *Allocation means the apportionment of a particular resource according to specific and pre-determined allocational principles.*

This rather broad definition points out two crucial aspects: First, we are dealing with a specific resource, which can be different in nature, e.g. financial, material, personnel or temporal. With respect to medical care, all of these examples play an important role. Secondly, its' apportionment is based upon specific and pre-determined principles to avoid any arbitrary and inefficient decision making.

Allocational resource *Allocational resource means any resource (e.g. financial, material, personnel or temporal), which by means of allocational principles is apportioned to indigent recipients.*

The specific character of allocational resources is that any apportionment is linked to allocational principles, which take into account an indigent recipient. The attempt of a principle-based apportionment of resources leads to the formulation of an allocational algorithm.

Allocational principles *Allocational principles refers to the criteria, which underlie the apportionment of resources.*

The definition of allocational principles is a crucial step to the formulation of an allocational algorithm. The character of allocational principles depends on their philosophical context. Liberalism, egalitarianism, utilitarism or contractualism set different priorities for ethically justifying their bases of allocation.

Allocational problem *Allocational problem means allocation under the pressure of limited resources, so that the number of indigent recipients exceeds the available resource.*

The term of allocational problem points out that allocation as defined above in general is not linked to the limitation of a resource. If the resource in question is limited, its apportionment is aggravated and needs further ethical justification.

Allocational ethics *Allocational ethics means the scientific reflection on the principle-based algorithm of allocation and its moral justifications.*

The term of allocational ethics underlines the importance of a reflection and justification of allocation by means of professional ethics. This is particularly true if allocational problems occur. The Challenges of allocation can arise within many different disciplines. Medicine is one such discipline, and as such, medical ethicists are obliged to deal with it.

6.3 Allocation, Allocational Problems and Tragic Choices

The term of allocation first came up in economic sciences and medical ethics undertook the term to characterize distributional problems in the context of transplant medicine (Schmidbauer 1974). The fact that the number of donor organs does not suffice to the need of transplant patients is a severe allocational problem. The work of, for instance, Eurotransplant is based upon a complex algorithm to alleviate these ethical dilemmas (Maio 2017, 434–435). It is also not difficult to discover further

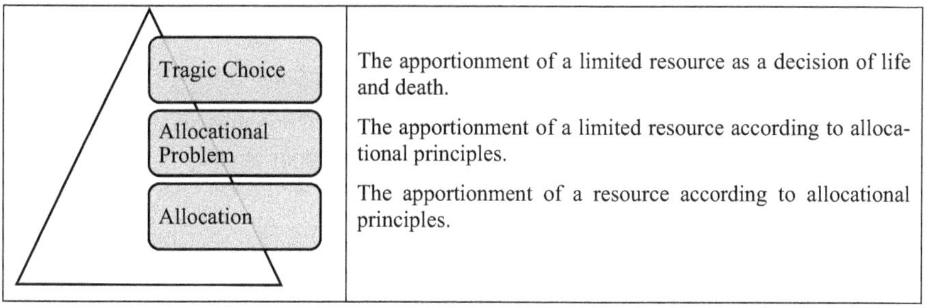

Fig. 6.1 Allocation, allocational problem and tragic choices

examples of allocational problems in medicine. With regard to military medicine, triage is one of the classic example (Rebentisch 2000; Gross 2006). The treatment of the civil population in countries of assignment is another one. An interesting historical example is the usage of penicillin during World War II (Porter 2000).

All of these examples have in common that the number of resources does not suffice to fulfill the urgent need of patients being dependent on them. As a result, it is impossible to avoid tragic choices, allowing some patients to survive while others cannot be saved. Whether in the context of civil or military medicine, the burdens which are necessary to make such decisions are heavy to bear, and all too often can lead to moral suffering or moral injury (Litz et al. 2009).

Allocation, allocational problems and the tragic consequences, require stringent medical ethical guidance. Particularly for young colleagues and comrades, the experience of allocational problems contradicts their self-understanding as medical professionals, who do not feel prepared to make a life or death decision. To prevent moral suffering or moral injury, it is most important to integrate the topic of allocation in medical ethical training (Fischer 2019). In a military context, a possible tragic choice due to allocational problems might become even more exacerbated in the light of dual loyalties (Fig. 6.1).

6.4 Implicit vs. Explicit Allocation

There is a general need to create awareness of allocational problems in relation to the professional field of activity. This is true for medicine and it is certainly true for military medicine. The fact that medical personnel might be confronted with very severe allocational dilemmas, should dictate that members of the medical profession (as well as other stakeholders in the military, in politics or society) know exactly how to deal with this problem. Depending on an implicit or an explicit proceeding, two forms of allocation can be distinguished: implicit allocation and explicit allocation (cf. Table 6.1).

Table 6.1 Explicit vs. Implicit allocation

Explicit allocation	Implicit allocation
Clear criteria of prioritization	Unclear criteria of prioritization
Open criteria of prioritization	Hidden criteria of prioritization
Clear intention to prioritize	Clear intention to prioritize

In the case of implicit allocation, some of the criteria of apportionment are not transparent. This is surly true in situations where allocation has not been reflected beforehand, and is actually being based upon arbitrary and opaque proceedings. This is also true, if those involved in an allocational process are not fully informed about the aims of allocation and the principles of allocation.

In contrast to this, explicit allocation fulfills the above made requirement of full awareness of allocation, allocational problems and allocational principles. In other words, those involved are aware of possible tragic choices and their consequences. Their personal decision is based on a fully transparent algorithm, which can therefore be justified not only to themselves, but also to both patients and to society at large.

6.5 Triage

As stated in the introduction, triage has today become a commonly known allocational algorithm widely discussed in society. The idea of triage first became urgent during the corona pandemic when intensive care capacities could no longer withstand the onslaught of patients, and physicians were forced to decide who received further treatment and who did not. Even long-serving emergency physicians were unprepared for this scenario. In addition to material and personnel bottlenecks (which hitherto had been typical characteristics of a mass-casualty incident), the long duration of the Covid pandemic, when coupled with the global reach of the pandemic were major additional considerations creating an entirely new state of affairs. It turned out that triage concepts from military medicine or disaster medicine, which help to handle temporally and spatially limited situations, could not easily be adopted to the situation of the pandemic (Hörnle et al. 2021).

Based on these early experiences, a number of allocational principles were developed, which might form the basis of triage as an allocational algorithm adopted to the covid-19-situation. It is quite interesting to see that the criteria discussed here were much more far-reaching than those generally adopted in military medicine or disaster medicine (Mannino 2021).

Triage in a military or disaster context has a strong utilitarian character based on the medical conditions of those competing for intensive care treatment. In fact, the utilitarian approach turned out to be the only efficient and effective proceeding in this context, at least when it came to saving as many lives as possible. In this context, the relevance of functional aspects regularly leads to controversial discussions,

Table 6.2 Triage related principles

Life & health care related principles	Economical related principles	Fate related principles
Age	Financial means	Luck
State of health	Health insurance	Lottery
Life expectancy	Patients benefit for society	
Medical history		

particularly in the case of reversal or wartime triage. The latter subordinates medical urgencies to military necessities and leads to severe problems of justification at least for deontologists in moral philosophy (Gross 2021). Isolated from the question of justification, triage as an allocational algorithm needs to be principle-based and wartime triage shows that these principles can be of a different nature.

Indeed, this plays an important role in the discussion about covid-19-related triage, which over the last year advanced as an example 'par excellence'. One does not have to be a prophet to predict that the triage during corona time will keep moral philosophers and medical ethicists busy for a long time thereafter. The main principles, which have been discussed in this context lead far beyond the discussion in military medical ethics (cf. Table 6.2). There are multiple criteria of significant importance when it comes to the formulation of allocational principles. Related to the physical condition of the patient principles like age, state of health, remaining life expectancy and individual medical history were also subjects to the discussion (life and health care related principles). Beside these considerations, principles like financial means, health insurance and patients benefit for society were brought up (economical related principles), as well as luck and lottery (fate related principles). Most interestingly classical bioethical principles and their underlying self-understanding of medical staff, patients and their relationship did not play a major role in this context. Though the principles like autonomy, non-maleficence, beneficence and justice are primarily applied to handle individual medical decisions, they surly play an important role in cases where a huge number of individuals request for medical help, too.

6.6 Allocational Problems in the Context of War

Questions concerning allocation and allocational problems have always been of primary concern to military medical personnel. They were surely at the forefront of Henri Dunant's (1828–1910) troubles when he was tasked with providing care for the wounded of Solferino. The distribution of penicillin to treat the sick and wounded in World War II provided a classic example of an allocation problem during a prolonged conflict. A more contemporary allocation problem might be reflected in the amount of military medical aid that could be provided to civilians in Afghanistan. Many times during the Afghan conflict, operational problems gave rise to open discussions about how best to implement personnel or distribute limited material resources.

Allocational problems may assume a morally questionable character if the decision at stake has to be based on the moral primacy of first fulfilling prioritised military and medical necessities. This can lead to tragic choices, which can cause moral stress to those who are entrusted to fulfil both the obligations of being a soldier and that of being a physician (dual loyalty conflict). An awareness of the meaning of allocation and the consequent challenge of allocational problems, is therefore of particular importance for military medical personnel.

6.7 Allocation and Bioethical Principles

It might surprise to look at allocational problems in the light of bioethical principles: autonomy, non-maleficence, beneficence and justice. Particularly when it comes to analyze triage, the overwhelming utilitarian nature of this algorithm seems to contradict the assumption that bioethical principles play a major role in this context. No doubt, the utilitarian approach with its focus on the benefit of the greatest number of patients seems to be most suitable to face allocational problems in medicine. Yet, the need of a corrective moment to soften the harshness of the utilitarian character and to prevent any functional abuse of triage has to be discussed. Reversal triage is just an obvious example for a functional manipulation, such as in the case of wartime triage where medical urgencies may become subordinate to military necessities.

Undoubtedly, during triage itself, medical personnel cannot consider and discuss the classical bioethical principles (autonomy, non-maleficence, beneficence and justice). Yet, this does not mean that they are irrelevant when it comes to the formulation of the triage algorithm. Bioethical principles focus on fundamental values, which are at stake in the case of allocational problem solving strategies. Especially in the case of triage, this becomes clear. The fundamental values, which are to be found at the basis of bioethical principles, are of enormous importance for the self-understanding of the physician, the self-understanding of the patient and the resulting physician-patient-relationship. They therefor need consideration both in advance and afterwards. By doing this, medical ethics takes into account the fate of the persons involved.

It is of crucial importance to contextualize triage in the broader context of the physician-patient-relationship. This claim might be controversial, as any moral overload should be avoided. Nevertheless, this step is necessary if physicians and medical staff, as well as patients and relatives, try to handle moral suffering resulting from a tragic triage decision. Disengagement surely helps to make the choices, which have to be made in the situation of triage. It is not an option when those who survived have to get on with their lives.

Table 6.3 Bioethical principles in the light of the Geneva declaration

Geneva declaration	Bioethical principle
"I will respect the autonomy and dignity of my patient"	The fact that autonomy goes together with dignity in this context underlines the enormous significance of the principle of ***autonomy***.
"The health and well-being of my patient will be my first consideration"	This passage strongly obliges the physician to ***beneficence***.
"I will maintain the utmost respect for human life"	Respect for human life recalls in the physician the principle of ***non-maleficence***.
"I will not permit considerations of age, disease or disability, creed, ethnic origin, gender, nationality, political affiliation, race, sexual orientation, social standing or any other factor to intervene between my duty and my patient	The principle of ***justice*** is based on an irrevocable obligation to anti-discrimination

6.8 The Physician-Patient-Relationship

The relationship between medical staff and patients has always been of the highest relevance. Classical texts of medical ethics like the Hippocratic Oath, as well as their modern versions underline its special importance. In our times, texts like the Geneva Declaration point out fundamental aspects, which demonstrate that the physician-patient-relationship is based on both obligation and trust. Although the oath as a whole is of vital consideration, in the light of allocational problems the following passages have a particularly outstanding character (cf. Table 6.3). They reflect the four principles of biomedical ethics, which cannot be disregarded with respect to the relationship between patients and healthcare-professionals.

The principles of biomedical ethics should permeate every physician-patient-relationship, regardless of whether the circumstance allow actual in depth reflection or not. This is particularly true for situations of allocational algorithms such as in the instance of triage. Intensive medical ethical training is essential to ensure that crucial principles like autonomy, beneficence, non-maleficence and justice remain the basis of any professional conduct.

6.9 Conclusion

Based on a set of definitions (allocation, allocational resource, allocational principle, allocational problem, allocational ethics) and a number of examples from civil and military medicine. I have highlighted out the need to create an awareness of allocation by means of a transparent and open conduct (explicit vs. implicit allocation). Triage in times of covid-19 has become a widely feared and discussed example for an allocational algorithm. The amount of allocational principles, which were discussed in this context, surmount by far the principles, which play an important role in military medicine. Although bioethical principles may appear to have previously been regarded to be of a negligible relevance concerning triage, I strongly advocate a more contextualized approach. As the physician-patient-relationship and its underlying ethos cannot refrain from bioethical principles, they should be considered particularly in the light of a tragic choice.

References

Fischer, Dirk. 2019. *Militärische Ethik im Militärischen Kontext. Eine Herausforderung für Forschung und Lehre.* In *Ethik und Militär. Kontroversen in Militärethik und Sicherheitspolitik* 2: 50–56.

Gross, Michael. 2006. *Bioethics and armed conflict. Moral dilemmas of medicine and war*, 18. Cambridge: MIT Press.

———. 2021. *Military medical ethics in contempory armed conflict. Mobilizing medicine in the pursuit of just war.* Vol. 31–33, 261–265. Oxford: University Press.

Höffe, Otfried. 2008. *Lexikon der Ethik*, 97. München: C. H. Beck.

Hörnle, Tatjana, et al. 2021. *Triage in der Pandemie*. Tübingen: Mohr Siebeck.

Litz, Brett, et al. 2009. Moral injury and moral repair in war veterans. A preliminary model and intervention strategy. *Clinical Psychology Review* 8: 695–706.

Maio, Giovanni. 2017. Mittelpunkt Mensch. In *Lehrbuch der Ethik in der Medizin*. Stuttgart: Schattauer.

Mannino, Adriano. 2021. *Wen rette ich – und wenn ja, wie viele? Über Triage und Verteilungsgerechtigkeit*. Stuttgart: Reclam.

Porter, Cole. 2000. *Die Kunst des Heilens. Eine medizinische Geschichte der Menschheit von der Antike bis heute*, 461. Heidelberg: Spektrum.

Rebentisch, Ernst. 2000. Geschichte der Sichtung. In *Udo Crespin and Gisela Neff*, ed. Handbuch der Sichtung, 31–40. Edewecht: Verlagsgesellschaft Stumpf & Kosssendey.

Schmidbauer, Herbert. 1974. *Allokation, technischer Fortschritt und Wettbewerbspolitik*. Tübingen: Mohr Siebeck.

Dirk Fischer Medical doctor, philosopher, and theologian. Doctor of medical history, doctor of moral theology, medical ethics consultant in the medical service of the Bundeswehr, head of the Teaching and Research Unit for Military Medical Ethics at the Bundeswehr Medical Academy Munich/Germany.

Chapter 7
Applying Battlefield Triage Ethics and Resource Allocation to a Contemporary Public Health Crisis: Lessons Learned from the Past and Adapting Them for the Future

Jeanne A. Krick, Jacob S. Hogue, Matthew A. Studer, Tyler R. Reese, and Elliott M. Weiss

Abstract As the world continues to battle the COVID-19 pandemic, the medical community faces the ongoing possibility of having to make difficult decisions regarding how to allocate scarce resources. Many groups have already developed triage frameworks to help guide healthcare teams in making these decisions. Those using and refining such algorithms can benefit from lessons learned in the long history of battlefield triage. This essay explores the ethical principles and priorities that guide decision-making during battlefield triage and compares them with those utilized during public health crises. We examine specific aspects of battlefield triage that can be translated to civilian public health crises, with recommendations on how

J. A. Krick (✉)
Department of Pediatrics, Brooke Army Medical Center, San Antonio, TX, USA
e-mail: Jeanne.a.krick.mil@mail.mil

J. S. Hogue
Department of Pediatrics, Madigan Army Medical Center, Tacoma, WA, USA
e-mail: jacob.s.hogue.mil@health.mil

M. A. Studer
Division of Cardiology, Department of Pediatrics, Seattle Children's Hospital, Seattle, WA, USA
e-mail: matthew.studer@seattlechildrens.org

T. R. Reese
Department of Family Medicine, Madigan Army Medical Center, Tacoma, WA, USA
e-mail: tyler.r.reese2.mil@health.mil

E. M. Weiss
Division of Neonatology, Department of Pediatrics, University of Washington, Seattle, WA, USA

Treuman Katz Center for Pediatric Bioethics, Seattle, WA, USA
e-mail: emweiss@uw.edu

S. M. Eagan, D. Messelken (eds.), *Resource Scarcity in Austere Environments*, Military and Humanitarian Health Ethics,
https://doi.org/10.1007/978-3-031-29059-6_7

to utilize them in these settings. Our hope is that health leaders during the current pandemic may draw upon important lessons learned over hundreds of years of battlefield triage when determining an ethical approach to managing scarce resources.

Keywords Triage · Resource allocation · Battlefield medicine · Military ethics

7.1 Introduction

Baron Dominique Jean Larrey, the surgeon general of Napolean's armies, revolutionized the care of soldiers on the battlefield through the innovative use of medical triage and evacuation of wounded personnel (Skandalakis et al. 2006). Part of his genius came in recognizing the importance of prioritization in treatment of patients, especially when resources were limited. His system of triage allowed resources to be primarily directed at returning soldiers to the battlefield, enabling the larger army to win battles and defeat their enemies. This use of triage significantly reduced the number of casualties in Napolean's army as compared with their opposing forces and was one of the many tactics that allowed for tremendous success in the early years of his campaigns.

Seeing the successful use of this battlefield triage and its impact on Napolean's ability to wage war, militaries across the world began adopting similar medical triage systems (Robertson-Steel 2006). These triage tactics adapted to changing battlefield conditions, centered on the idea of utilizing limited medical resources in a way that best preserves the fighting force to complete its primary military objective. In modern military medicine, healthcare providers are taught the principles of medical triage early in their careers and frequently train on its use during military training exercises and operational planning.

As successful medical triage and patient evacuation evolved on the battlefield, civilian providers began applying these lessons to prehospital and emergency care. Following the Vietnam War, triage evolved within the United States to prioritize the care of the most critically ill and injured for evacuation and care at the nearest hospital (Mitchell 2008). Over time, this was adapted and expanded to civilian disaster settings, where the number and acuity of patients overran the medical capabilities and resources available both at the scene and nearest emergency departments (e.g. Las Vegas shooting, September 11th attacks) (Shackelford et al. 2022; Melmer et al. 2019; Menes et al. 2017; Feeney et al. 2005). Medical triage is now routinely incorporated into settings where prioritization of patients is necessary to maximize lifesaving measures while preserving material and human resources (Nakao et al. 2017). In typical civilian settings where resources are not limited, the most critical patients are prioritized for care and given nearly-unlimited resources. When resources are limited in these disaster settings, triage priorities shift towards preserving scarce resources for those patients most likely to benefit from their use.

The COVID-19 pandemic has similarly resulted in circumstances where patient needs have stressed health systems' capacities to the point of collapse. There has been a relentless drum of critically ill patients and high volume surges which have overwhelmed hospital capacities, requiring leaders to consider and implement crisis standards of care (Alaska 2021). In crisis standards of care, the needs of the patient community overwhelm the hospital capacity to the point where it becomes necessary to alter the usual standards of care in order to optimize the use of scarce resources for the benefit of an entire community (Committee on Guidance for Establishing Crisis Standards of Care for Use in Disaster Situations; Institute of Medicine 2012). In these circumstances, medical organizations have had to establish COVID-19-specific triage guidelines with the goal is to most effectively use scarce resources to preserve the most lives possible. In developing these guidelines, community and hospital leaders have relied on principles of triage, including the ethical justifications of such triage, to appropriately determine not only who receives limited resources (e.g. ventilators, ICU beds), but also in planning for the distribution of novel therapeutics and vaccines (Herlitz et al. 2021).

The ethical principles central to the practice of battlefield triage are not dissimilar to those utilized in developing civilian triage priorities, with a few notable exceptions. Civilian community and medical leaders can gain valuable lessons from examining and understanding the military's long experience with battlefield triage, to include the ethical principles tantamount to such deliberation, planning and execution in the setting of public health crises. In this chapter, we will examine the ethical principles underlying battlefield triage and consider how they may inform practices during a public health crisis.

7.2 Ethical Foundations

To better understand the ethical underpinnings of triage, it is important to first examine some guiding bioethical principles. Within the field of medicine, ethical practice is commonly guided by a principlist approach, which centers on the four basic ethical principles of beneficence, non-maleficence, justice, and respect for autonomy (Beauchamp and Childress 2019). These principles offer an analytic framework for medical decision making and serve as a starting point for biomedical ethics. The principle of respect for autonomy is often granted the highest priority in Western medicine, where the focus of care is on the wishes of an individual patient (Childress 1990). This individual patient, when well-informed, is typically granted the latitude to make autonomous decisions regarding their medical care, even if their choices do not correspond with the recommendations of the medical team, require high resource utilization, or have very little chance of achieve a desired outcome. Aside from limitations imposed by insurance or other financial constraints, this is the decision-making environment most familiar to those living within the United States.

When medical resources become scare, priorities may mandate a shift away from the individual and redistribute the weight of the ethical principle of respect for autonomy towards one of justice and a focus on the community. Providing the greatest good to the greatest number of patients within a community, even at the expense of individual medical liberties, may have to become the "new standard." (Emanuel et al. 2020) This more strict utilitarian approach to care during times of limited resources focuses on the aggregate good of the group and gives no weight to the needs of individual patients (Rawls 1971).

7.3 Ethical Priorities on the Battlefield

For those within the military medical community, balancing the ethics of respect for autonomy (i.e. a focus on the individual patient) and justice (i.e. a focus on the community) is familiar and changes depending on the circumstance of where they are providing care. During times of war, medical providers are guided by the ethical principles set out during the Geneva Conventions of 1949 (International Committee of the Red Cross 1950). These conventions set out priorities for the protection of the wounded and sick that have formed the basis of international humanitarian law during conflicts (Schussler et al. 2019). During times of war, triage officers are appointed to make decisions that align with the larger military mission, at times subverting the preferences of individual patients. These appointed officers must understand and abide by the ethical principles of the Geneva Conventions in determining priorities of care. While these triage officers are generally physicians, other medical personnel including dentists, nurses and medics have sometimes served this role (Janousek et al. 1999).

In the setting of a mass casualty event on the battlefield, triage officers must consider not only an individual patient's physical injuries, but also the larger military mission, availability of resources at their site, evacuation capabilities, and likelihood of successful treatment of the individual patient (Beam 2003). The ethical principle of military necessity offers justification for this practice, where preferential care may be given to individuals who are most able to aid in the war effort (Gross 2021; Gross 2017). When triage officers encounter a patient during an event with multiple injured military members and the devastating injuries make survival unlikely, they are obligated by this same principle to designate them to a category where death is imminent, known as "expectant," in order to preserve scarce resources (e.g. blood products) for those who are most likely to survive and potentially return to duty (Repine et al. 2005). This stands in stark contrast with patients receiving typical care in a civilian setting, where the same patient may consume a significant amount of material resources and manpower before ultimately succumbing to their injuries. The larger military mission drives the ethical priorities in these battlefield encounters.

In a battlefield setting, scarce resources can be prioritized based on the status of individuals. Most often this is seen where medical rules of eligibility determine if care can be provided to civilians within a host nation. When resources became more scarce during recent conflicts in Iraq and Afghanistan, local civilians only became eligible for medical care when their injuries were a direct result of Coalition action and there was space available at the Coalition medical facility (Gross 2021; Gross 2017; Woll and Brisson 2013). The principle of military necessity again justifies this practice, in that scarce resources are prioritized for those serving the primary military mission. When resources are more available, however, civilians may be treated by these same facilities, especially if a larger component of the military mission is to gain trust with the local population. This was seen throughout different time periods during the recent wars in Iraq and Afghanistan (Schauer et al. 2018; Naylor et al. 2020; Gurney et al. 2022).

In extraordinary circumstances, this same principle of military necessity may shift triage priorities to give precedence to friendly combatants with minor injuries (over friendly or enemy combatants with more severe injuries) who may be quickly returned to the fight. This "reverse" or "salvage" triage is often justified from an ethical lens in that it may allow for a decrease in the overall number of casualties on the battlefield by bringing a swifter end to the battle with a larger fighting force (Gross 2008; Edwards 2009). The utilitarian argument calculates the greatest good as the greatest number of lives saved above all else, including lives on both sides of the battlefield. It is important to note that, while this practice has been justified from an ethical perspective, it is not routinely practiced.

While the Geneva Conventions specifically bar triage based on identity (i.e. rank, country of origin, etc.), there are times where it may be ethically justified to provide differential care to local nationals, both soldier and civilian, on the battlefield (International Committee of the Red Cross 1950). The principle of medical necessity allows for this disparate treatment for similar wounds when external factors influence the success of treatment (Gross 2021; Gross 2017). An equivalent application of this principle in the civilian sector is in the determination of organ transplant recipients. Factors external to an individual's health, such as housing, access to follow-up, and reliability of the patient to adhere to treatments all influence their ability to be listed for organ transplantation given the scarcity of viable organs. In recent military conflicts in Iraq and Afghanistan, host-nation allies were at times offered a different standard of care as compared to similarly wounded Coalition soldiers based on factors such as access to long-term care and the chances of treatment success. An example of this is in the treatment of soldiers with extensive burns in Afghanistan. While Coalition soldiers could be repatriated to their home nations with burn units to care for their injuries, doctors were instructed not to intubate local Afghans with burns exceeding 50% of their body, as they would not have access to high-quality burn care and were not likely to survive, despite the availability of resources (Sokol 2011).

When individuals volunteer to join the U.S. military, they choose to become a part of an organization that achieves its goals both implicitly and explicitly through commitment to the Constitution and service to a greater good. By virtue of service, a military member has agreed to suspend some individual liberties in favor of national priorities. The need to shift to a community-focused medical ethic in the context of a military massive casualty event aligns with this intrinsic mindset of military service. While the circumstances creating the need for medical triage on the battlefield are often chaotic, the principles are generally understood and accepted by those who will fall under its practice. For those in the military, this allows for a degree of trust in the triage officer to apply fair and equitable standards when making these difficult decisions to maximize the benefit of scarce resources. These individuals may be more willing to sacrifice their personal autonomy in these dire situations, especially if they have been trained on the relevant military medical ethical principles and have practiced their application in a training environment.

7.4 Public Health Ethics and Triage

During a public health crisis, the prognostication of survival is of utmost importance in determining who should receive limited resources. Public health and health system officials have an obligation to safeguard resources in such a manner that they are utilized to maximize benefits for the most patients (Emanuel et al. 2020; Domres et al. 2010). In theory, this would obligate that life-saving resources be prioritized for patients with the greatest likelihood of survival. While logical and straightforward in theory, the practical application of this principle is fraught with difficulty. There is an inherent challenge in attempting to predict a patient's likelihood of recovery based on objective factors at the time of presentation to the medical system (Vasilevskis et al. 2016; Wynants et al. 2020; Aziz et al. 2020). These predictions are complex and involve wide ranges of age, innumerable sets of comorbidities, and evolving understanding of the presenting disease or trauma. Then within that prognostic milieu, determining whether long- or short-term survival should be prioritized also remains controversial (Joebges and Biller-Andorno 2020; Sprung et al. 2020). Additionally, it is almost impossible to predict how long a patient will require a particular resource, such as a ventilator. This becomes especially problematic when a patient, who is already using a limited resource, no longer has the highest likelihood of benefit from its use, but its discontinuation would result in the patient's death.

In addition to safeguarding resources in a manner that maximizes the medical benefit to the greatest number of patients, triage during a public health crisis ought to make efforts to overcome existing and emerging health disparities (Baker and Strosberg 1992). This stems from an egalitarian view of triage, where the assumption is that there is equal value to all human life, everyone has an equal right to healthcare, and everyone has equal access to the point of care (Beauchamp and Childress 2019). Based on the initial conditions of equal rights and human value, those determining the allocation of scarce resources during a public health crisis have a more complex responsibility to ensure they are connected with the local community, addressing barriers that may prevent some from access to those resources (Schmidt 2020; Ballantyne et al. 2020; Ballantyne 2020).

Unlike military service members who train with and generally accept battlefield triage, community members in a public health crisis have likely never experienced such rationing and may not be as accepting. Knowing this, public health leaders must create triage standards that will be acceptable to their local communities or risk considerable resistance from those within them. To develop this public trust, transparent and engaging processes must include the community in developing such guidelines for resource allocation.

Both prior to and throughout the COVID-19 pandemic, several communities engaged in this type of transparent and cooperative policy development work for their local medical systems (Daugherty Biddison et al. 2019; Biddison et al. 2018; Archer et al. 2020; Marshall et al. 2021). Despite this evidence of collaboration, it is unclear how community members responded to such triage guidelines, especially within the United States where patients are not accustomed to being restricted from life-prolonging resources. While imperfect, it is important for public health experts to involve community members in the development of any triage guidelines and to publicize the ethical foundations behind such guidelines as widely as possible to aid in the acceptance within a community.

7.5 Lessons from the Battlefield

Communities across the world have faced extreme resource challenges throughout the COVID-19 pandemic. Many countries have been forced to impose strict criteria for the use of life-saving resources (e.g. ventilators, ICU beds) due to an overwhelming number of patients with critical medical need (Mounk 2020). Some have left rationing decisions to individual hospital systems or communities, often with guidance from large medical organizations (Vergano et al. 2020; White and Lo 2020; Berlinger et al. 2020). Healthcare workers have struggled to implement bedside rationing decisions, as it is in direct conflict with the basic tenet of providing care for the sick and dying (Davidson et al. 2020). These frontline healthcare workers are vulnerable to moral injury and there has been widespread moral distress across healthcare (Lamiani et al. 2021; Lemmo et al. 2022; Vig 2022; Alonso-Prieto et al. 2022; Borges et al. 2020). This moral distress and residue is not unlike that experienced by medical workers on the battlefield who experience massive casualty events (Wilson 2021). While the military is far from perfect in addressing such distress, civilian medical systems may benefit from examining the lessons that the military has learned over the years in the importance of attending to such suffering.

Beyond the distress of caring for sick patients during a public health crisis, healthcare workers can experience heightened feelings of anguish when they are responsible for triage decisions. While those caring for individual patients will have the most insight into their clinical course and most-likely outcome, they have an inherent obligation to that individual patient. On the battlefield, there is a distinct role of a triage officer, who assigns the order in which patients are treated based on the accepted triage principles of the military. They must factor in the available resources, a simplified needs assessment of the patient, and the overall military mission. This assigned individual, who must have some medical knowledge, is not typically part of the primary treating team and may not even be a nurse or physician (Janousek et al. 1999; Repine et al. 2005). This allows the medical treatment teams to focus on their primary responsibilities to the individual patients they see before them, creating a separation between them and the distress of having to make difficult triage decisions. While the battlefield is not perfectly analogous to the resource limitations during a public health crisis, hospitals may assign triage officers (or teams) that are responsible for making the difficult decisions of resource allocation, based on accepted guidelines. Many institutions have created such triage committees that are responsible for making decisions about rationing of care for individual patients (Naidoo and Naidoo 2021; Jöbges et al. 2020). These teams may help alleviate the burden of moral injury felt by healthcare workers throughout such a crisis.

While most ethical guidance emphasizes the importance of transparency in the development of any rationing or triage principles, individual hospitals and hospital systems have struggled with the process of making their guidelines known to their community of potential patients. Even regional systems that have well established guidelines that are easily available to the public (e.g. through public websites) can expect some degree of surprise and frustration from patients and families when typical critical resources are not available under such rationing decisions. As has been discussed, there is inherent difficulty in getting a community to accept such a major change to their expectations of healthcare. While it is important for community members to serve as representative voices in the development of any triage and rationing guidelines, that only begins the process of trust-building between the local community and the medical system. For it to mature, the process of triage and framework for how decisions are made must be well-publicized throughout a community. With that, a community may be able to appreciate the value in consistent medical decision making, even if they are not satisfied with an individual decision.

On the battlefield, this is accomplished in part through standard military trainings, where all service members learn these principles long before they encounter such situations. While there is no perfect parallel within the civilian sector, medical systems can ensure that their employees are well-versed in protocolized decision making processes during crisis situations. Health system employees must be afforded the opportunity to engage with leadership in understanding their main priorities during triage, much as soldiers on the battlefield must understand the main priority of accomplishing a military mission. Additionally, these health systems and public health entities should make guidelines regarding the process of decision making as publicized within their community as possible. Hospital leadership may feel apprehensive about publishing their guidance given the controversial nature of the topic, but transparency and familiarization with triage standards prior to critical resource limitation is an exercise in building trust with both their employees and surrounding community. It is only through this establishment of trust that there can be a community-minded approach to resource allocation.

Lastly, there may be utility in considering the battlefield concept of "salvage" or "reverse" triage when developing triage priorities during a public health crisis

(Kearns 2021). During such a crisis, skilled healthcare providers are some of the most scarce resources and there are some who argue that these individuals should get priority for some treatments and protective measures, as their return to health will allow them to take care of other patients in need. The use of scarce resources on healthcare workers may provide a "multiplier effect," in that it would allow for this care of future patients. The prioritization of healthcare workers in this manner may be justified given their instrumental value to society during a public health crisis (Real de Asua and Fins 2022). It is important to acknowledge that the application of this concept may only apply to some resources, as healthcare workers needing more intensive therapies may not be able to return to work and provide care to other patients. In these cases, the ethical justification of an increased instrumental value of healthcare workers would be insufficient and it would not be appropriate to give priority to these individuals.

7.6 Conclusion

There are many differences between the triage practiced on a battlefield and that required during a public health crisis. Some of these differences arise from the underlying ethical principles that take priority in the different environments, while others are due to the overarching goals in the two settings. Importantly, the distinct nature of the two populations (i.e. military and civilian) necessitate different strategies for the planning, implementation, and practice of triage medicine. Those within the military accept that the suspension of personal liberties for the sake of a larger mission may be necessary and may be more accepting of utilitarian minded triage strategies on the battlefield. Civilian populations during a public health crisis, however, will be much less accustomed to such limitations. They may have difficulty placing their trust in a larger medical system to determine which patients receive critical resources, especially if there is a history of structural inequities across healthcare. Still, public health officials can learn from the hundreds of years of military triage on the battlefield and may benefit from applying these lessons in a way that translates to their local community.

Disclaimer Statement The views expressed herein are those of the authors and do not reflect the official policy or position of Brooke Army Medical Center, the U.S. Army Medical Department, the Defense Health Agency, the U.S. Army Office of the Surgeon General, the Department of the Army, the Department of the Air Force, or the Department of Defense or the U.S. Government.

References

Alaska, Soucheray S. 2021. Idaho using crisis standards of care over COVID-19. *CIDRAP News* [Internet]. 15 September 2021. Available from: https://www.cidrap.umn.edu/news-perspective/2021/09/alaska-idaho-using-crisis-standards-care-over-covid-19.

Alonso-Prieto, E., H. Longstaff, A. Black, and A.K. Virani. 2022. COVID-19 outbreak: Understanding moral-distress experiences faced by healthcare Workers in British Columbia, Canada. *International Journal of Environmental Research and Public Health* 19 (15).

Archer, R.A., A.I. Marshall, K. Sirison, W. Witthayapipopsakul, P. Sriakkpokin, S. Chotchoungchatchai, et al. 2020. Prioritizing critical-care resources in response to COVID-19: Lessons from the development of Thailand's Triage protocol. *International Journal of Technology Assessment in Health Care* 36 (6): 540–544.

Aziz, S., Y.M. Arabi, W. Alhazzani, L. Evans, G. Citerio, K. Fischkoff, et al. 2020. Managing ICU surge during the COVID-19 crisis: Rapid guidelines. *Intensive Care Medicine* 46 (7): 1303–1325.

Baker, R., and M. Strosberg. 1992. Triage and equality: An historical reassessment of utilitarian analyses of triage. *Kennedy Institute of Ethics Journal* 2 (2): 103–123.

Ballantyne, A. 2020. ICU triage: How many lives or whose lives? *BMJ Journal of Medical Ethics Blog* [Internet] 7 April 2020. Available from: https://blogs.bmj.com/medical-ethics/2020/04/07/icu-triage-how-many-lives-or-whose-lives/.

Ballantyne, A., W.A. Rogers, V. Entwistle, and C. Towns. 2020. Revisiting the equity debate in COVID-19: ICU is no panacea. *Journal of Medical Ethics* 46 (10): 641–645.

Beam, T.E. 2003. Medical ethics on the battlefield: The crucible of military medical ethics. In *Military medical ethics*, ed. T.E. Beam and L.R. Sparacino, vol. 2. Washington, D.C.: TMM Publications.

Beauchamp, T.L., and J.F. Childress. 2019. *Principles of biomedical ethics*. 8th ed. New York: Oxford University Press.

Berlinger, N., Wynia, M., Powell, T., Hester, D.M., Milliken, A., Fabi, R., et al. 2020. Ethical framework for health care institutions and guidelines for institutional ethics services responding to the coronavirus pandemic. The Hastings Center [Internet] 16 March 2020. Available from: https://www.thehastingscenter.org/ethicalframeworkcovid19/

Biddison, E.L.D., H.S. Gwon, M. Schoch-Spana, A.C. Regenberg, C. Juliano, R.R. Faden, et al. 2018. Scarce resource allocation during disasters: A mixed-method community engagement study. *Chest* 153 (1): 187–195.

Borges, L.M., S.M. Barnes, J.K. Farnsworth, N.H. Bahraini, and L.A. Brenner. 2020. A commentary on moral injury among health care providers during the COVID-19 pandemic. *Psychological Trauma* 12 (S1): S138–SS40.

Childress, J.F. 1990. The place of autonomy in bioethics. *The Hastings Center Report* 20 (1): 12–17.

Committee on Guidance for Establishing Crisis Standards of Care for Use in Disaster Situations; Institute of Medicine. 2012. *Crisis standards of care: A systems framework for catastrophic disaster response*. Washington (DC): National Academies Press (US); 21 March 2012.

Daugherty Biddison, E.L., R. Faden, H.S. Gwon, D.P. Mareiniss, A.C. Regenberg, M. Schoch-Spana, et al. 2019. Too many patients…a framework to guide statewide allocation of scarce mechanical ventilation during disasters. *Chest* 155 (4): 848–854.

Davidson, P.M., W.V. Padula, J. Daly, and D. Jackson. 2020. Moral outrage in COVID19-understandable but not a strategy. *Journal of Clinical Nursing* 29 (19–20): 3600–3602.

Domres, B., T. Kees, S. Gromer, P. Braitmaier, and T. Granzow. 2010. Ethical aspects of triage. *Rocnik* 2: 76–82.

Edwards, M. 2009. Triage. *Lancet* 373 (9674): 1515.

Emanuel, E.J., G. Persad, R. Upshur, B. Thome, M. Parker, A. Glickman, et al. 2020. Fair allocation of scarce medical resources in the time of Covid-19. *The New England Journal of Medicine* 382 (21): 2049–2055.

Feeney, J.M., R. Goldberg, J.A. Blumenthal, and M.K. Wallack. 2005. September 11, 2001, revisited: A review of the data. *Archives of Surgery* 140 (11): 1068–1073.

Gross, M.L. 2008. Why treat the wounded? Warrior care, military salvage, and national health. *The American Journal of Bioethics* 8 (2): 3–12.

———. 2017. Saving life, limb, and eyesight: Assessing the medical rules of eligibility during armed conflict. *The American Journal of Bioethics* 17 (10): 40–52.

Gross, M. 2021. *Military medical ethics in contemporary armed conflict: Mobilizing medicine in the pursuit of just war*, 92–111. New York, NY: Oxford University Press.
Gurney, J.M., V. Graf, A.M. Staudt, J.D. Trevino, C.A. VanFosson, H. Wild, et al. 2022. Characterization of humanitarian trauma care by US military facilities during combat operations in Afghanistan and Iraq. *Annals of Surgery* 276: 732–742.
Herlitz, A., Z. Lederman, J. Miller, M. Fleurbaey, S. Venkatapuram, C. Atuire, et al. 2021. Just allocation of COVID-19 vaccines. *BMJ Global Health* 6 (2): e004812.
International Committee of the Red Cross. 1950. *The Geneva Conventions of August 12, 1949 analysis for the use of national Red Cross societies*. Geneva: International Committee of the Red Cross. Available from: http://www.loc.gov/rr/frd/Military_Law/RC_Gen-Conv_Analysis.html
Janousek, J.T., D.E. Jackson, R.A. De Lorenzo, and M. Coppola. 1999. Mass casualty triage knowledge of military medical personnel. *Military Medicine* 164 (5): 332–335.
Jöbges, S., R. Vinay, V.A. Luyckx, and N. Biller-Andorno. 2020. Recommendations on COVID-19 triage: International comparison and ethical analysis. *Bioethics* 34 (9): 948–959.
Joebges, S., and N. Biller-Andorno. 2020. Ethics guidelines on COVID-19 triage-an emerging international consensus. *Critical Care* 24 (1): 201.
Kearns, A.J. 2021. The principle of salvage in the context of COVID-19. *Nursing Inquiry* 28 (1): e12389.
Lamiani, G., D. Biscardi, E.C. Meyer, A. Giannini, and E. Vegni. 2021. Moral distress trajectories of physicians 1 year after the COVID-19 outbreak: A grounded theory study. *International Journal of Environmental Research and Public Health* 18 (24).
Lemmo, D., R. Vitale, C. Girardi, R. Salsano, and E. Auriemma. 2022. Moral distress events and emotional trajectories in nursing narratives during the COVID-19 pandemic. *International Journal of Environmental Research and Public Health* 19 (14).
Marshall, A.I., R. Archer, W. Witthayapipopsakul, K. Sirison, S. Chotchoungchatchai, P. Sriakkpokin, et al. 2021. Developing a Thai national critical care allocation guideline during the COVID-19 pandemic: A rapid review and stakeholder consultation. *Health Research Policy and Systems* 19 (1): 47.
Melmer, P., M. Carlin, C.A. Castater, D. Koganti, S.D. Hurst, B.M. Tracy, et al. 2019. Mass casualty shootings and emergency preparedness: A multidisciplinary approach for an unpredictable event. *Journal of Multidisciplinary Healthcare* 12: 1013–1021.
Menes, K., J. Tintinalli, and L. Plaster. 2017. *How One Las Vegas ED Saved Hundreds of Lives After the Worst Mass Shooting in U.S. History*. Emergency Physicians Monthly [Internet]. 3 November 2017. Available from: https://epmonthly.com/article/not-heroes-wear-capes-one-las-vegas-ed-saved-hundreds-lives-worst-mass-shooting-u-s-history/
Mitchell, G.W. 2008. A brief history of triage. *Disaster Medicine and Public Health Preparedness* 2 (Suppl 1): S4–S7.
Mounk, Y. 2020. The extraordinary decisions facing Italian doctors. *The Atlantic* [Internet] 11 March 2020. Available from: https://www.theatlantic.com/ideas/archive/2020/03/who-gets-hospital-bed/607807/
Naidoo, R., and K. Naidoo. 2021. Prioritising 'already-scarce' intensive care unit resources in the midst of COVID-19: A call for regional triage committees in South Africa. *BMC Medical Ethics* 22 (1): 28.
Nakao, H., I. Ukai, and J. Kotani. 2017. A review of the history of the origin of triage from a disaster medicine perspective. *Acute Medicine & Surgery* 4 (4): 379–384.
Naylor, J.F., M.D. April, E.E. Thronson, G.J. Hill, and S.G. Schauer. 2020. U.S. Military medical evacuation and prehospital Care of Pediatric Trauma Casualties in Iraq and Afghanistan. *Prehospital Emergency Care* 24 (2): 265–272.
Rawls, J. 1971. *A theory of justice*. Vol. xv, 607. Cambridge: Belknap Press of Harvard University Press.
Real de Asua, D., and J.J. Fins. 2022. Should healthcare workers be prioritised during the COVID-19 pandemic? A view from Madrid and New York. *Journal of Medical Ethics* 48 (6): 397–400.

Repine, T.B., P. Lisagor, and D.J. Cohen. 2005. The dynamics and ethics of triage: Rationing care in hard times. *Military Medicine* 170 (6): 505–509.
Robertson-Steel, I. 2006. Evolution of triage systems. *Emergency Medicine Journal* 23 (2): 154–155.
Schauer, S.G., G.J. Hill, J.F. Naylor, M.D. April, M. Borgman, and V.S. Bebarta. 2018. Emergency department resuscitation of pediatric trauma patients in Iraq and Afghanistan. *The American Journal of Emergency Medicine* 36 (9): 1540–1544.
Schmidt, H. 2020. The way we ration ventilators is biased. *New York Times* [Internet]. 15 April 2020. Available from: https://www.nytimes.com/2020/04/15/opinion/covid-ventilator-rationing-blacks.html
Schussler, L., F.M. Burkle, and S.M. Wren. 2019. Protecting surgeons and patients during wars and armed conflicts: Importance of Predeployment training on the Geneva conventions and international humanitarian law. *JAMA Surgery* 154 (8): 683–684.
Shackelford, S.A., M.A. Remley, S. Keenan, R.S. Kotwal, J.B. Baker, J. Gurney, et al. 2022. Evidence-based principles of time, triage and treatment: Refining the initial medical response to massive casualty incidents. *Journal of Trauma and Acute Care Surgery* 93 (2S Suppl 1): S160–S1S4.
Skandalakis, P.N., P. Lainas, O. Zoras, J.E. Skandalakis, and P. Mirilas. 2006. "To afford the wounded speedy assistance": Dominique Jean Larrey and Napoleon. *World Journal of Surgery* 30 (8): 1392–1399.
Sokol, D.K. 2011. The medical ethics of the battlefield. *BMJ* 343: d3877.
Sprung, C.L., G.M. Joynt, M.D. Christian, R.D. Truog, J. Rello, and J.L. Nates. 2020. Adult ICU triage during the coronavirus disease 2019 pandemic: Who will live and who will die? Recommendations to improve survival. *Critical Care Medicine* 48 (8): 1196–1202.
Vasilevskis, E.E., P.P. Pandharipande, A.J. Graves, A. Shintani, R. Tsuruta, E.W. Ely, et al. 2016. Validity of a modified sequential organ failure assessment score using the Richmond agitation-sedation scale. *Critical Care Medicine* 44 (1): 138–146.
Vergano, M., G. Bertolini, A. Giannini, G.R. Gristina, S. Livigni, G. Mistraletti, et al. 2020. Clinical ethics recommendations for the allocation of intensive care treatments in exceptional, resource-limited circumstances: The Italian perspective during the COVID-19 epidemic. *Critical Care* 24 (1): 165.
Vig, E.K. 2022. As the pandemic recedes, will moral distress continue to surge? *The American Journal of Hospice & Palliative Care* 39 (4): 401–405.
White, D.B., and B. Lo. 2020. A framework for rationing ventilators and critical care beds during the COVID-19 pandemic. *Journal of the American Medical Association* 323 (18): 1773–1774.
Wilson, M.A. 2021. A tool adaptation workshop for moral distress in military critical care nurses. *Clinical Nurse Specialist* 35 (4): 208–214.
Woll, M., and P. Brisson. 2013. Humanitarian care by a forward surgical team in Afghanistan. *Military Medicine* 178 (4): 385–388.
Wynants, L., B. Van Calster, G.S. Collins, R.D. Riley, G. Heinze, E. Schuit, et al. 2020. Prediction models for diagnosis and prognosis of covid-19 infection: Systematic review and critical appraisal. *BMJ* 369: m1328.

Dr. Jeanne A. Krick is an Active-Duty U.S. Army Neonatologist and Ethicist, stationed in San Antonio, Texas. She currently serves as the Consultant to the Surgeon General of the Army for Medical Ethics. She has experience as an ethics consultant in both military and civilian hospital systems and is a current member of the Texas Ethics Consortium. She has research interests in military medical ethics, parent-physician communication in the Neonatal Intensive Care Unit, and periviability. She is married to Joseph, a former Army armor officer, and they have one son, Alex.

Dr. Jacob S. Hogue is an Active-Duty Army Geneticist, stationed in Tacoma, Washington. He currently serves as the chair of the Ethics Committee at Madigan Army Medical Center.

Dr. Matthew A. Studer is a retired Army Pediatric Cardiologist, who previously served as Ethics leadership at multiple military treatment facilities. He was deployed multiple times during his military career. He currently works in Seattle, Washington.

Dr. Tyler R. Reese is an Active-Duty Army Family Medicine and Palliative Care Physician. He has previously chaired the Ethics committee at Madigan Army Medical Center and currently serves as the Deputy Consultant to the Army Surgeon General for Medical Ethics.

Dr. Elliott M. Weiss is a Neonatologist and Ethicist who works in Seattle, Washington. His research interests include informed consent, shared decision-making, and neonatal bioethics.

Chapter 8
Medical Triage by Moral Responsibility in Crisis and War

Stephen N. Woodside

Abstract International Humanitarian Law mandates that all wounded in war, no matter which party they belong to, shall receive aid in accordance with their medical condition, and that "there shall be no distinction among them founded on any grounds other than medical ones." This principle of impartiality is endorsed by various other military and civilian institutions worldwide to include the ICRC, the US Department of Defense, and the American Medical Association. In this essay, I argue that in some cases, we ought, morally speaking, to triage based on factors beyond one's medical condition. More specifically, I argue that in these cases, we ought to triage, at least partly, based on one's moral responsibility for one's current medical predicament. I call this view Triage by Moral Responsibility (TMR). I support my argument for TMR by demonstrating its plausibility when applied to cases of medical treatment and disaster relief, and by drawing parallels between these situations and those considered in the ethics of defensive harming. I conclude by discussing the practicality of my moral conclusion for the practice of medical triage in pandemics and war.

Keywords Harm · Liability · Medical triage · Moral responsibility · Pandemic · War

8.1 Introduction

The problem I am addressing has to do with medical triage in what I am going to henceforth call *limited resource situations* (LRSs). These are situations in which we cannot meet the medical needs of all patients or victims due to a scarcity of resources

S. N. Woodside (✉)
United States Military Academy, West Point, NY, USA
e-mail: stephen.woodside@westpoint.edu

S. M. Eagan, D. Messelken (eds.), *Resource Scarcity in Austere Environments*, Military and Humanitarian Health Ethics,
https://doi.org/10.1007/978-3-031-29059-6_8

such as medical personnel, medicine, equipment, or time. LRSs in the most extreme cases are forced-choice situations where we must allow one or more persons to die in order to save another. Here is the kind of case I have in mind:

> *Lung Transplant*: Two patients, Hank and Mandy, are both in need of a lung transplant to survive, but only one lung is available. Hank has smoked for the last 30 years despite being reasonably informed of the dangers of smoking. He knew many years ago that he ought to have quit, but he never felt that it was worth the effort. Mandy, on the other hand, just had bad luck. She has never smoked or been responsible for other risky behavior that might have contributed to her condition. They both have the same good prognosis if given the new lung, both would die were they not to receive the transplant, and we can assume that this situation will be enough to scare Hank off from ever touching another cigarette.

We can state the LRS problem as the following: How morally ought[1] we to medically triage in LRSs? Whom ought we to save, and whom ought we allow to die when we cannot save them all? The LRS problem is fundamentally a question of distributive justice and promoting the good. The question, then, can be more precisely formulated as the following: How ought we to medically triage in LRSs in a way that promotes the most good while treating potential victims fairly?

To be clear, I am framing the LRS problem as a moral one, and so my main argument and conclusion will be moral ones. Of course, the LRS problem is also a legal one, and perhaps also a professional one in terms of what codes of conduct various medical and humanitarian organizations should adopt and follow. But while my main argument reaches only a moral conclusion, I am also going to suggest that it does give us good reason to reconsider various legal and professional codes regarding medical triage in LRSs in domestic crises and war. I will come back to some of these codes shortly.

One attractive moral principle regarding medical treatment in general is that it is morally impermissible to triage by factors other than a patient's medical condition. I will refer to this principle as *Impartiality*:

> **Impartiality**: We ought to triage only on the basis of one's medical condition. Triage through consideration of anything else (e.g. race, sex, nationality) is morally impermissible.

This is a plausible moral view, which is reflected in various legal and professional codes worldwide. Nevertheless, I am going to argue that Impartiality is false. In doing so, I am going to argue that, other things being equal, we ought to prioritize those who are less morally responsible for their current medical predicament in LRSs. A corollary of my thesis is that we ought to allow those who are more morally responsible for their condition to die when we cannot save everyone. I call this view *Triage by Moral Responsibility*:

> **Triage by Moral Responsibility (TMR)**: Other things being equal, we ought to prioritize treatment of those who are less morally responsible than others for their current medical

[1] I will sometimes drop 'morally' for sake of brevity. Other uses of 'ought' will be explicitly stated as such. The same is true for other normative terms such as 'permission,' 'wrong,' 'responsibility,' 'blame,' etc.

predicament in LRSs. We ought to save those who are less morally responsible and allow those who are more responsible to die when we cannot save them all.

Moral responsibility is a scalar concept; a person can be more or less responsible for her current condition, she can be minimally or maximally responsible for it, or she can lack any responsibility whatsoever for it. Moral responsibility for one's condition is a function of, among other things perhaps, the degree to which one's actions or omissions[2] contributed to one's condition, the moral status of those actions, and any excusing conditions that might have been present, for example duress, coercion, or nonculpable ignorance. If one is morally responsible for some action or resulting state of affairs, it does not follow that one is morally blameworthy (or praiseworthy) for it (McMahan 2009, 162). While moral responsibility is a necessary condition for blameworthiness, appropriate blame seems to require a sufficient degree of moral responsibility. One could be morally responsible to some degree for wrongfully harming another, but if this was the result of coercion or acting from non-culpable ignorance, they might not be an appropriate target of blame. Thus, one can be morally responsible without being blameworthy.

It is important to clarify that my answer to the LRS problem is only a partial solution. That is, TMR claims that moral responsibility is at least one factor by which we ought to medically triage in LRSs, which is consistent with the claim that other factors ought to be considered as well. Once again, it is the claim that, other things being equal,[3] we ought to save those who are less morally responsible for their current predicament and allow the more responsible others to die.

On the face of it, TMR might sound harsh or punitive. I will dispute the latter characterization later in the paper. As for its harshness, if by that one means 'unfair', I will also dispute that. Indeed, I will argue that fairness *requires* adhering to TMR. Another potentially unsettling feature of TMR is that it flies in the face of the orthodox view, Impartiality. But, as I will first argue, Impartiality is morally problematic. Before I begin with my discussion of and objections to Impartiality though, I want to lay out some assumptions that I will not argue for. These assumptions seem fair to me, as they seem to be also endorsed by proponents of Impartiality. Many of these assumptions would, however, be rejected by pure consequentialists, but they are not my intended audience. I want to convince those who acknowledge the existence of deontological moral rights that TMR respects these rights, while Impartiality does not.

---

[2] I will hereafter simply use 'actions' to include omissions.

[3] My 'other things being equal' clause is meant to hold constant the good that would be produced by treating one vs. another. This would include, among other things, that medical need, effectiveness of treatment, and the number and value of future years of life are the same across patients.

8.2 Some Assumptions

My first assumption, which I have just alluded to, is that the correct solution to the LRS problem must consider both the moral value of outcomes and respect for moral rights. That is, while the goodness of outcomes is a morally relevant factor in triage, we must not triage in a way that violates the rights of those affected. This means, of course, that in some cases, what we ought to do in a particular LRS will not be the option that produces the most good.

Second, I assume that all persons have the right of fair access to required medical care, but that this right can be justifiably overridden when enough good is at stake. This is known as a *moderate* deontological position, which rejects the *absolutist* view that moral rights can never be justifiably overridden.[4] This latter extreme position essentially holds that we ought to respect moral rights no matter the cost. The distinction between the two views can be easily seen in considering the moral right of free speech, assuming there is such a right. My right to free speech imposes a constraint on others to not deny me that right, even if doing so would promote more good in certain cases. For example, suppose I were delivering this paper as a public address to some crowd. Imagine that some in the crowd would be angered by my remarks and that, as a result, slightly more good would come about by my being silenced. In this case, both the moderate and absolutist positions would agree that it would be wrong to silence me, even though doing so would produce more net good. But suppose instead that I were to deliver my address at a worldwide medical conference. Imagine that in this case my remarks would set off a massive strike by medical workers across the world, resulting in millions of preventable deaths. While the absolutist position would still say that it would be wrong to silence me, the moderate one might say that I ought to be silenced here, even though doing so would infringe on my right to free speech. It might be that *I* am wronged in being silenced here, and consequently that I am owed some compensation (perhaps an apology at least), but all things considered, silencing me is not wrong. By my lights, this is the more plausible of the two deontological positions, but I will not argue any further for it.

Lastly, I assume that factors such as race, nationality, sex, and the like are not morally relevant factors for medical triage.

8.3 Impartiality

Impartiality is the view with which most of us are familiar, and it also seems to be the view that is endorsed by most medical and humanitarian organizations worldwide. Again, we can state it as the following:

---

[4] I follow Kagan (1997, Ch. 3) in my discussion of this distinction.

> **Impartiality**: We ought to triage only on the basis of one's medical condition. Triage through consideration of anything else (e.g. race, sex, rationality) is morally impermissible.

To clarify, Impartiality is not simply claiming that one's medical condition is *a* factor by which we ought to triage; it says it is the *only* morally permissible factor. It is worth noting that there is some plausibility and attractiveness to Impartiality, but I will ultimately argue that it is false. Before I argue this though, I want to first note several endorsements of the principle along with some related rationale for them.

8.3.1 Endorsements of Impartiality

The first endorsement can be found in International Humanitarian Law (IHL), which states the following:

> The wounded, sick and shipwrecked must receive, to the fullest extent practicable and with the least possible delay, the medical care and attention required by their condition. No distinction may be made among them founded on any grounds other than medical ones (ICRC 2021).

Of course, this is a *legal* mandate, and so it need not commit one to the *moral* principle of Impartiality. However, if Impartiality as a moral principle is false, this might give us good reason to question IHL here. I will come back to this point later in the paper.

This endorsement of Impartiality comes from the International Committee of the Red Cross (ICRC):

> [ICRC] makes no discrimination as to nationality, race, religious beliefs, class, or political opinions. It endeavors to relieve the suffering of individuals, being guided solely by their needs, and to give priority to the most urgent cases of distress (ICRC 2015).

The ICRC goes on to say this and its six other principles "sum up the movement's ethics" and that they are "key to the movement's identity" (2015). This suggests that the ICRC sees Impartiality not only as a legal requirement (under IHL), but also as an ethical code of conduct for its workers. It is interesting to note here that the ICRC also mentions some factors by which workers should not triage—nationality, race, religious beliefs, class, or political opinions. That seems superfluous given the claim that workers ought to triage only by medical need. However, it might suggest their motivation for endorsing Impartiality. Medical triage that takes into consideration factors such as nationality, race, religion, class, or political affiliation is obviously unfair. This is obvious, to most of us at least, because these are some of the most egregious examples where the right to fair access is violated. There are other obvious examples such as triage that uses gender or sexual orientation to discriminate. Considering all these obvious violations of fair access and our desire to eliminate them, one might be tempted to draw the conclusion that we ought to triage based only on medical condition. This motivation for Impartiality is understandable, but it runs the risk of ignoring factors that might be morally relevant in determining fair access.

Here is an endorsement by the American Medical Association (AMA):

> Differences in treatment that are not directly related to differences in individual patients' clinical needs or preferences constitute inappropriate variations in health care […] [P]hysicians […] are called on to provide the same quality of care to all patients without regard to medically irrelevant personal characteristics (AMA 2021b).

And like the ICRC, the AMA sees this as a sound ethical principle; it is listed as one of their "Principles of Medical Ethics" (2021).

The United States Department of Defense endorses Impartiality with the following:

> Discrimination [regarding health care] on the basis of age, sexual orientation, gender, race, ethnicity, language, disease, disability, religion, or rank is forbidden because it is inconsistent with the ideals and principles of the [Military Health System] (US DoD 2017).

Interestingly, this endorsement, like the ICRC's, mentions various factors by which not to triage. However, perhaps as an improvement on the ICRC's statement, it does not rule out other potentially permissible factors by which to triage. Perhaps this is motivated by military reasons, or perhaps there is some suggestion here that there are other morally relevant factors by which to triage, as I will argue there are.

Lastly, consider this additional statement from the AMA:

> The commitment to fair access to care runs throughout the AMA Code of Medical Ethics […] Opinion 1.1.2, "Prospective Patients," instructs physicians to uphold ethical responsibilities not to discriminate on the basis of "personal or social characteristics that are not clinically relevant to the individual's care," […] and Opinion 11.1.4, "Financial Barrier's to Health Care Access," define health care as "a fundamental human good" that entails obligations to promote access to care on the part of individual physicians, the medical profession, and society at large (AMA 2021a).

We can summarize the AMA's rationale in the following way: We all have the right to fair access to required medical care, and following Impartiality is the only way to respect this right. And it is worth emphasizing once again that the AMA views Impartiality not simply as a legal or professional obligation, but also as an ethical one.

Piecing all these preceding endorsements together, we can begin to see an argument for Impartiality based on promoting the good and respecting rights. The argument is not that we should promote the most good without concern for the rights of individuals. If that were the rationale, then why would some of these codes explicitly forbid triage by race, nationality, or gender? Again, these kinds of triage would already be prohibited by stating that only triage by medical condition is permissible. These types of triage are mentioned because they are obviously unfair methods. They deny people the right to required medical care based on morally irrelevant grounds. Thus, these proponents of Impartiality are concerned about respecting the rights of individuals, as they should be. The mistake, though, is in thinking that the rights of individuals are always respected when we triage only by medical need. It is a false assumption that *all* grounds beyond one's medical need are morally irrelevant, so I will argue.

8.3.2 Against Impartiality

One obvious problem with Impartiality is that it is morally silent in some LRSs. If Impartiality is true and two or more victims have the same medical need, then whom ought we to save, and whom ought we allow to die? Clearly, we ought to save somebody, but how ought we to choose according to Impartiality? One possible answer is that it would be permissible to save either. Given that both have the same medical need, it would be permissible to save either and let the other die. It is simply left to the preference of the medical professional. But can that be right? Is that really the fair way to choose? I suspect not.

Perhaps in recognition of this morally problematic implication, the AMA offers this guidance in the type of LRS I have in mind:

> Procedurally fair criteria for allocating limited resources base decisions first on medical need and likelihood of benefit (and then, when criteria of medical need do not substantially distinguish among patients), an objective mechanism like a lottery. Decisions to allocate scarce resources should not be based on non-medical criteria, such as judgments about a patient's "social worth," or the degree to which the patient may have contributed to illness (AMA 2021a).

Admittedly, this AMA guidance is preferable to a decision-making procedure based on personal preference. It might also provide some comfort to medical workers charged with the difficult decision of whom to treat in LRSs. If the ultimate decision is left up to chance, then perhaps the physician is likely to feel that the decision is not *her* decision. Thus, she might feel less guilt in being involved in a situation where one is allowed to suffer harm. But is this a fair result? Let us revisit the case we began with:

> *Lung Transplant*: Two patients, Hank and Mandy, are both in need of a lung transplant to survive, but only one lung is available. Hank has smoked for the last 30 years despite being reasonably informed of the dangers of smoking. He knew many years ago that he ought to have quit, but he never felt that it was worth the effort. Mandy, on the other hand, just had bad luck. She has never smoked or been responsible for other risky behavior that might have contributed to her condition. They both have the same good prognosis if given the new lung, both would die were they not to receive the transplant, and we can assume that this situation will be enough to scare Hank off from ever touching another cigarette.

Suppose as the AMA suggests, the doctor decides to flip a coin to decide, and it goes in Hank's favor. He gets to live, while Mandy must die. Is that the fair outcome? One might argue that it is by insisting that Hank and Mandy had an equal *chance* of getting the lung. That is certainly true, but equality of chance does not imply fairness. Imagine a case where a father leaves it up to his twin daughter and son whether to help him with some necessary lawn work. It is clear to both that he could really use the help, but the father would prefer to let them use their best judgment regarding what to do. As it turns out, the daughter decides to help her father, while the son decides to play video games. Well into their lawn work, the father comes upon a silver dollar partially buried in the grass. He knows that both the twins would love to have it, as it would make an excellent addition to each of their coin collections. Assuming he is going to give it to one of them, whom ought he give it to? The father knows, of course, that if he decides to give it to the daughter, his son will be upset. His son might even call it unfair. But I take it that, to most of us, this would be a dubious claim. Indeed, an unfair outcome would be for the son to get it, even if by a coin toss.

To be clear, I am not claiming that it would be unfair for Hank to get the lung or for the son to get the silver dollar simply because the other is denied it. Being granted fair access to something is consistent with its being ultimately denied. Again, the son should not cry "foul" when the coin is given to his sister since she seems to have a stronger claim toward it. Rather, fair access would be violated only if the reasons for denial were either based on morally irrelevant factors or did not account for factors that were relevant. Triage based on race, sex, etc. are examples of the former kind of unfairness, while the lung going to Hank (or the coin to the son) are examples of the latter kind.

All this is not to say that tie-breaking decision-making procedures such as a coin toss or lottery are inherently unfair. However, they would be fair only in cases where there is an actual *tie*. The relevant sense of 'tie' for our purposes is that two potential victims are identical with regard to all morally relevant factors. And that is simply not the case in *Lung Transplant*.

In summary, if Impartiality is true, then it is morally permissible to give the lung to Hank rather than giving it to Mandy. But this is not the case, as it violates her right of fair access to required medical care. Thus, Impartiality is false.

8.4 TMR, Liability, and the Just Distribution of Harm

Lung Transplant not only serves as a counterexample to Impartiality, but it also suggests that TMR is true, as this seems to be the best explanation for why Mandy ought to get the lung. Here is the view once again:

> **Triage by Moral Responsibility (TMR)**: Other things being equal, we ought to prioritize treatment of those who are less morally responsible than others for their current medical predicament in LRSs. We ought to save those who are less morally responsible and allow those who are more responsible to die when we cannot save them all.

I now want to further support the argument for TMR by drawing some parallels between it and the concept of moral liability to defensive killing, which has been best defended by Jeff McMahan (2005, 2009). While LRSs are not cases involving *defensive* harming, they, like cases of defensive harming, are fundamentally settled by questions regarding the "just distribution of harm" (McMahan 2005, 395).

8.4.1 Moral Liability to Defensive Killing

The concept of moral liability provides a nice rights-based explanation for the permissibility of harming in self-defense. In the usual circumstances, I have the right not to be harmed by another. However, if I threaten a person with wrongful harm, I can forfeit that right, such that it would be permissible for him to harm me in self-defense. I can become morally liable to his defensive harm. Because I am liable to the harm, I am not wronged by it, as I no longer have a right against it. I have forfeited this right. This explains why defensive harming is often morally permissible.

McMahan (2005) has argued that what makes someone liable to defensive harming is moral responsibility for a threat of wrongful of harm. Applying this to our above case, I forfeit my right against being harmed by my victim by virtue of my being morally responsible for my wrongful threat to him. This is not necessarily to say that I deserve to be harmed by him. On McMahan's view, one can be liable to harm without deserving it. All that is required on his view is moral responsibility, which need not rise to a level of culpability that might be required for deserved harm. Here is one of McMahan's cases that demonstrates this distinction:

> *The Resident*: The identical twin of a notorious mass murderer is driving in the middle of a stormy night in a remote area when his car breaks down. He is nonculpably unaware that his twin brother, the murderer, has within the past few hours escaped from prison in just this area, and that the residents have been warned of the escape. The murderer's notoriety derives from his invariable modus operandi: he violently breaks into people's homes and kills them instantly. As the twin whose car has broken down approaches a house to request to use the telephone, the resident of the house takes aim to shoot him, preemptively, believing him to be the murderer (2009, 164).

On McMahan's view, although the resident is not culpable to a degree where she deserves harm, she is nevertheless liable to the twin's defensive harm by virtue of her moral responsibility, however minimal it might be, for the wrongful threat to him.

McMahan (2005) goes on to support his view on liability by appealing to the concept of distributive justice. On his view, liability is fundamentally a matter of the just (or fair) distribution of harm in situations where harm is inevitable. In the *Resident*, it is certain that someone will be killed. Either the resident will kill the twin, or the twin can kill the resident in self-defense. Given that the resident is more responsible than the twin for the situation in which someone will die (since she is responsible for the wrongful threat to begin with), it is fair that she should suffer the harm. In other words, she is liable to the twin's defensive harm, which means that, in this case, she is liable to be killed by him.

Again, the claim is not that the resident ought to be killed because she *deserves* it. If that were true, then the twin would have reason to kill the resident even absent any instrumental considerations such as saving his own life. That seems doubtful, however. It seems that in this case, the twin is morally permitted to kill the resident only because it is a situation of inevitable harm. And the resident is liable to the harm since she is more responsible than the twin for this situation.

8.4.2 LRSs and Liability to Allowed Harm

Like situations involving defensive harm, LRSs are cases where it is inevitable that someone must suffer harm. Fundamentally, then, they are dilemmas regarding liability to harm. They are fundamentally matters of distributive justice. So, if moral responsibility for the situation of inevitable harm settles questions of liability to defensive harm, it should also settle questions of liability to harm in LRSs.

Reconsidering *Lung Transplant*, this suggests that Hank, rather than Mandy, ought to suffer the harm by being denied the lung. While Hank is not responsible for the threat to Mandy *per se* (her illness is responsible for that), he is responsible, and more so than Mandy, for the situation in which someone must suffer harm. In other words, he is more responsible than Mandy for their being involved in an LRS where one must die for the other to live. Thus, as a matter of the just distribution of harm, it seems fair that Mandy should have the lung and Hank should die. Hank is liable to the harm.

Let us consider another LRS, this time involving a natural disaster:

> *Rescue*: A category-5 hurricane is approaching the Outer Banks (North Carolina) and based on tracking data, the residents of Hatteras (southern tip) are given a strict evacuation order 48 hours out, while those of Corolla (northern tip) are told it will be safe to shelter in place and weather the storm. As it turns out, both communities are devastated by the storm. As the storm subsides, two same-sized families, one each from Hatteras and Corolla, are clinging to survival on the tops of their houses as the water continues to rise. Only one of the families can be evacuated by a helicopter in time. The other will die.

Let us assume that the Hatteras family was properly informed of the evacuation order and had ample opportunity to evacuate prior to the storm. They just could not be bothered by the evacuation. We will also assume that both rescue operations have an equally good chance of success. Who ought to be saved? The first thing to say is that this will likely be an emotionally difficult decision for whoever has to make it. If only there were a way to save both families, that is what we would and should do. And it is certainly a bad thing that one of them will die. They do not deserve that. Nevertheless, one family must die, so who should it be?

My answer, at this point, should be obvious. If we think that this is fundamentally a matter of fairness regarding the distribution of harm, then Hatteras ought to suffer it. They are liable to suffer the harm. Of course, we might reach a different verdict regarding liability if Hatteras had some morally compelling reason to remain at their home despite the evacuation order. For example, suppose Hatteras had a family member who could not evacuate due to serious medical reasons. In that case, it might seem that Hatteras is no more responsible than Corolla for the LRS. And in that case, perhaps a coin toss would be a fair method to decide who will suffer the harm. But again, there is no moral tie in the case as originally conceived. So, Hatteras is liable to suffer the harm. That is the fair, albeit regrettable, result.

There is, however, at least one important difference between LRSs such as *Lung Transplant* and *Rescue* and cases of defensive harming such as the *Resident*, and it is a difference that speaks in favor of my view. While cases of self-defense are ones in which it is inevitable that harm will be *done* to someone, LRSs are ones in which we must *allow* someone to be harmed. Many moral philosophers, at least those who acknowledge the existence of deontological moral rights, believe that there is a morally significant difference between doing and allowing harm, such that the former is more difficult to justify than the latter (Quinn 1989; Woollard 2015).[5] For example, while it might take the prevention of a lethal threat of harm to justify killing (i.e. doing harm), the justification for allowing death might only require the prevention of, say, a lost limb. This suggests that it will take substantially less good (or bad to be prevented) to justify allowing death in LRSs than it does to justify killing in cases of self-defense. So, if one is inclined to accept McMahan's view on liability to defensive harm, one should be even more eager to accept TMR.

8.5 Some Objections and Replies

I now want to turn to some possible objections to my view. Some of these objections target my moral thesis, while others target the practical importance of it, even if it is true.

Objection 1: TMR denies some people the right to required medical care, and on a moderate deontological view, this denial would be all-things-considered permissible only if the value of following TMR vastly outweighed the value of not doing so. But that seems implausible. Indeed, it is likely that value of following TMR would be worse than not following it and adhering instead to Impartiality.

> *Reply*: It is obviously true that TMR implies that some ought to be denied required medical care in LRSs. But I do not believe that access to required medical care is a moral right.[6] Again, I believe that all have the right of *fair access* to required medical care, which can be respected even in cases where access is denied. The right of fair access is potentially violated when one is denied medical care based on morally irrelevant factors (e.g. race or religion) or based on a lack of consideration of factors that are morally relevant (e.g. medical need or moral responsibility). As for the claim that the consequences of following TMR would be worse than those of following Impartiality, I am happy to concede that for the sake of argument. For even if that is true, it does not follow that Impartiality is true and TMR is false. While the value of outcomes is a morally relevant factor for both views, it is not the only factor. Both views recognize that we should seek to maximize the goodness of outcomes, but only in a way that is consistent with respecting moral rights. This leaves open the possibility that the morally right way to triage would not produce the best consequences.

Objection 2: TMR seems to suggest that people should be punished disproportionately (e.g. death in some cases) for their comparatively minor wrongdoings (e.g. smoking). But no one deserves to die for such things!

---

[5] See Rachels (1975) for one of the most compelling arguments against this moral distinction.

[6] See Thomson (1971) for a nice argument that the moral right to life does not imply a right to all those things necessary for living.

Reply: While I agree that a punishment of death for the wrongdoers I have discussed would be disproportionate, and hence morally unjustified, TMR is grounded on distributive, rather than retributive justice. That is, while I agree that the wrongdoers discussed do not deserve to die, it is fair that they, rather than another, die. This is not punishment. To clarify, if it were possible to save all victims or patients, then we ought to do that on my view. A retributivist view might suggest otherwise.

Objection 3: One of the most important things about Impartiality is that it builds trust between our healthcare workers and relief organizations and the public. Following TMR would erode this trust, which would result in disastrous consequences. Avoiding this is more important than fair access.

Reply: I am sympathetic to this objection. If following TMR were to have the consequences suggested, this might give us grounds to justifiably infringe one's right to fair access. Again, on my view, fair access to required medical care, like all other deontological rights, is not an absolute right. When enough good is at stake, it can be all-things-considered permissible to deny fair access. So even if TMR is true, it would not have much practical importance since, according to this objection, practitioners ought not to follow it due to the disastrous consequences of doing so. But this objection rests on an empirical claim, which, at the very least, deserves further investigation. I agree that trust between practitioners and the public is critically important for effective healthcare, but triage by moral responsibility need not degrade this critical element. For one, transparency would seem to be necessary for continued trust, which suggests that medical professional codes ought to be revised to account for TMR.

Objection 4: Your view assumes that healthcare workers will know the comparative levels of moral responsibility of their patients. Obviously, this will often be impossible to determine.

Reply: No, TMR does not assume this, although I am somewhat sympathetic to the underlying concern. Admittedly, determining victims' levels of moral responsibility would be tricky business in many cases, particularly those in an emergency room setting. However, moral responsibility does seem to be more determinable in more deliberate situations such as *Lung Transplant* and *Rescue*. In those cases, which are not so fantastical, it seems pretty clear who has more responsibility for the LRS in which one must die.

Objection 5: Perhaps you are correct that, morally speaking, moral responsibility is one factor by which we ought to triage in LRSs. And while that might be of interest to moral philosophers, it is simply not a realistic proposal for doing actual triage. Along with the epistemic concerns already stated, practitioners, who operate in the real world, need to be concerned with legal liability for their choices. And triaging with any consideration of moral responsibility is likely to put them in a precarious legal situation.

Reply: I concede that following TMR is not advisable for the legal reasons given, at least given the current legal situation. But if TMR is true, then it does give us good reason to reconsider current domestic and international law regarding medical triage to see how they might better reflect our moral principles. Furthermore, even if we ultimately decide that the laws are all-things-considered best as written, we should not pretend that Impartiality is a sound ethical principle. In writing this, I am reminded of a similar point made by James Rachels (1975). In arguing against the prevailing view that there is a morally significant difference between active and passive euthanasia, which is supposed to support various medical ethics codes strictly prohibiting active euthanasia but permitting passive euthanasia in some cases, he makes the following point:

> [D]octors may think that all of this is only of academic interest—the sort of thing that philosophers may worry about but has no practical bearing on their own work. After all, doctors must be concerned about the legal consequences of what they do, and active euthanasia is clearly forbidden by the law. But even so, doctors should also be concerned with the fact that the law is forcing upon them a moral doctrine that may well be indefensible, and has considerable effect on their practices…So, whereas doctors may have to discriminate between active and passive euthanasia to satisfy the law, they should not do any more than that. In particular, they should not give the distinction any added authority and weight by writing it into official statements of medical ethics (Rachels 1975, 80).

Similarly, even if medical professionals are required by law to follow Impartiality, organizations such as the AMA should not give it any added value by including it as part of their ethical code and suggesting that it is "fair."

8.6 TMR, COVID-19, and Vaccination

Some of the most contentious questions regarding the current COVID-19 pandemic are those concerned with, broadly speaking, the ethics of vaccination. Since pandemics seem to necessarily involve LRSs, at least in their initial stages, TMR certainly commits me to certain positions on these questions. In this section, I am going to focus on two of these questions and suggest what TMR might say about them. The questions are: (1) Assuming individuals have a moral right to refuse vaccination, ought the fact that one exercises this right count in favor of prioritizing care for another otherwise similar vaccinated person? And (2) how ought we to prioritize vaccinations themselves in LRSs?

8.6.1 The Moral Right to Refuse Vaccination

It does seem likely that some individuals have the right to refuse vaccination in cases where they have morally compelling reasons to do so. The most obvious cases are those involving persons who would risk being seriously harmed by the vaccine due to, say, underlying medical conditions. The less obvious ones involve those who refuse vaccination for religious reasons. But it seems less likely that those who refuse vaccination out of stubbornness or unfounded skepticism have this right. This would amount to saying that they have the right to risk harming other persons for no good reason. Nevertheless, for the sake of argument, I am going to assume that all individuals have the right to refuse vaccination no matter their reasons for doing so. What follows from this? Does it follow that their refusal to get vaccinated ought not to be potentially used against them if involved in a COVID-related LRS? No. Even assuming that all have the right to refuse vaccination, this does not entail that it is unethical to use their refusal as a reason to prioritize others in LRSs. Nevertheless, there is an argument to be considered for the opposing view.

One might argue that if it is permissible for one to refuse vaccination whatever one's reasons, then they cannot be held responsible for the consequences of their permissible action. So, if one of the consequences happens to be that they become involved in an LRS, their refusal to vaccinate ought not to be used as a reason to treat others over them. For this would be to hold them responsible for the outcome of something they had every right to do.

This is a valid argument, but it is unsound. The key premise of the argument seems committed to the following more general principle: *if one is morally permitted to do x, then one is not morally responsible for any of the consequences of x.* This principle is clearly false. While it is true that there are some risk-imposing actions that are morally permissible, we can still be morally responsible for their consequences, particularly if we *foresee* the possibility of their occurring. Many actions in war result in foreseeable yet unintentional harm to innocent civilians, and many agree that these actions can be morally permissible provided certain other conditions.[7] And while it would seem unreasonable to blame or punish the agents for the foreseeable harm, since their actions were permissible, it does seem reasonable to suggest that they are nevertheless morally responsible to some degree for it. Indeed, they might be responsible to such a degree that they incur an obligation to compensate. Thus, it seems that one can be morally responsible for the harmful consequences of a permissible action.

Of course, we have been assuming, for sake of argument, that those who refuse vaccination for no good reason are permitted to do so. But there is even less to say in their defense when it comes to matters of responsibility for the foreseeable risk of harm, either to themselves or others, compared with combatants who might have morally compelling reasons to inflict foreseeable harm on civilians. If I am correct about all this and TMR is true, then refusal to vaccinate could be morally relevant grounds by which to triage, even assuming we all have the right to refuse. And if that is the practice by which medical professionals not only ought to, but also *will* triage, then perhaps the public is owed some transparency by virtue of a clearly stated policy on such. This would likely have the additional benefit of increased vaccination rates.

8.6.2 Vaccination Priority

As in simpler LRSs such as *Lung Transplant* and *Rescue*, government and medical officials have sought to develop vaccination strategies with an eye toward fairness and efficiency. This seems appropriate since, as I have argued, LRSs are

---

[7] According to the doctrine of "double effect" (Walzer 1977, Ch. 9), one of these conditions is that the good effects must be proportional to (i.e. outweigh) the bad effects. This is known as proportionality.

fundamentally problems of promoting the most good, in this case preventing illness, within the constraints of fairness. In an effort to satisfy this tricky pair of criteria, many of them have prioritized front-line health care workers, first responders, and those with "high-risk medical conditions" that make them more susceptible to the virus. This last category includes, among others, the obese and smokers. Some have questioned the fairness of this policy on grounds similar to the argument for TMR. But even if TMR is true, the argument against this policy is not that straightforward.

First, to reiterate, TMR does not claim that moral responsibility is the *only* criterion by which we ought to triage. It is simply one criterion, albeit an important one on my view. That medical need is also a weighty criterion for morally sound triage is consistent with TMR. And it is reasonable to assume that those with high-risk medical conditions, even those who are at fault for them, have substantially greater medical need than others. Thus, perhaps they ought to be prioritized in being protected from the virus. On the other hand, it might be that in arguing for TMR, I am committed to rejecting the prioritization of these individuals. Suppose that a high-risk smoker has a 30% chance of dying from COVID-19 if she is not immediately vaccinated, and suppose that I, a non-high-risk person, has only a .05% chance of dying if vaccinated at some later time.[8] Even if that is true, is it fair that I should incur the risk, no matter how small, rather than the irresponsible smoker incurring the greater risk? If not, then unless the good to be achieved by infringing my right to fair access were sufficiently large, it would not be permissible to do so.[9]

These are complicated issues, and I am unsure of the correct answers. While it does seem unfair that I should incur even a much smaller risk to mitigate the smoker's much larger one, the good produced by doing so is difficult to determine. There are at least two ways to consider the matter. First, we might think that the risk of harm, in itself, has negative value, while the prevention of it has positive value. And so, the net good of vaccinating the smoker would be substantially greater than that of vaccinating me (.05% vs. 30% chance of someone dying). And perhaps that is enough to justify infringing my right to fair access. Alternatively, we might think that the good (or bad) is only to be found in the consequences. And while the risk has no value in itself, it is the best we can do to predict the consequences and their value. And so, while it might end up being the case that more good would have resulted from my being vaccinated before the smoker (suppose she would have died either way, while I would die just in case I am not vaccinated first), we simply could not know that. Thus, a policy that prioritizes based partly on risk of harm seems understandable, even if it ends up resulting in right violations. In other words, perhaps we should not blame those responsible for such a policy.

---

[8] These percentages are obviously not accurate, but they should still serve to make my point.

[9] Recall that on a moderate deontological view, the right to fair access can be permissibly infringed when enough good is at stake.

8.7 TMR and the Medical Ethics of War

War is one of those calamities in which we are likely to find an abundance of LRSs. While the resource of medical supplies might not always be an issue, the mass casualties often associated with war greatly constrain the resource of time. Very often, we simply do not have enough time to save all the victims of war. So, our now familiar LRS question reappears: Whom ought we to aide and whom ought we allow to suffer, and perhaps die, when we cannot treat them all? Whom ought to receive priority of medical care and aid from organizations such as the ICRC when they cannot relieve the suffering of all? In other words, who is morally liable to suffer the harm?

Traditional accounts of the ethics of war have argued that whether one is morally liable to harm in war depends on the type of group one with which one is associated (Walzer 1977). On this view, all combatants are morally liable to harm since they belong to a group that is inherently threatening, while all noncombatants are not. The business of noncombatants is peace, not war. This, of course, is the view endorsed by IHL, which permits the targeting of combatants from both sides of a conflict and prohibits targeting all noncombatants.

Turning to the question of medical treatment and aid in war, the traditional view would seem committed to prioritizing noncombatants over combatants for the same reasons as above. If the forced choice is between saving a noncombatant or a combatant, both with lethal wounds, we ought to prioritize the former since they are not liable to even *allowed* harm, while the latter are.[10]

Over the last two decades, however, the traditional view has come under fire by rival revisionist views such as "reductive individualism" (Frowe 2015). On this view, the moral principles governing war are fundamentally no different from those in domestic situations such as individual self-defense. Thus, what makes one liable to harm in war is no different from what makes one liable in individual cases of self-defense—moral responsibility for a threat of wrongful harm (McMahan 2009). Consequently, most just combatants (i.e. those fighting for the just side of a war) and most noncombatants are not liable to be killed, while most unjust combatants are liable. For the former are generally not responsible for a threat of *wrongful* harm, while the latter are. So, if reductive individualism is true, it seems that TMR would hold even in wartime LRSs. And if that is true, there would seem to be no moral reason to prioritize the treatment of noncombatants over just combatants since neither is morally responsible for a threat of *wrongful* harm. It is, of course, true that just combatants are morally responsible for harming, but their harming is, generally speaking, justified.[11] It is the unjust combatants who are morally responsible for the threat of wrongful harm, whereby noncombatants and just combatants

---

[10] Recall my discussion of doing vs. allowing harm in Sect. 8.4.2.

[11] There are exceptions to this, of course. Those who fight on the just side of war can commit war crimes or other morally impermissible actions, which could make them liable to defensive harm and allowed harm in LRSs. I am concerned with the more controversial question of whether just

find themselves in need of treatment. Thus, TMR seems to imply that noncombatants should not receive priority of medical care over just combatants.

Admittedly, this would be a somewhat unsettling implication of TMR, but I do not believe that TMR actually implies it. Indeed, I think it is consistent with TMR that treatment of civilians should be prioritized over that of just combatants, at least in many cases. Consider the case of a just combatant who permissibly attacks her unjust counterpart, which results in the foreseeable yet unintentional injuring of a nearby innocent civilian. Suppose that our just combatant is also injured to roughly the same extent by enemy counter-fire. It is consistent with TMR that we ought to prioritize treatment of the civilian. While it is true that our just combatant has injured the civilian through, by hypothesis, morally permissible action, that she foreseeably harmed a non-liable person makes her, to some degree, morally responsible for it.[12] Thus, she is more morally responsible than the civilian for the LRS, which, according to TMR, gives us reason to save the civilian over her.

But what about cases in which neither combatants nor civilians have any moral responsibility for an LRS, or at least have the same level of responsibility for it? Imagine a case where a group of just combatants are injured by a roadside bomb, which also injures a similar number of civilians. Whom ought to be prioritized? This is a more difficult case, at least if one is a reductive individualist. But even here, there is still an argument to be made that it is consistent with TMR that we ought to prioritize the civilians. We might say here that even though both the just combatants and the civilians have an equal moral claim to be treated, or that it is fair that we treat either, we still have more reason to treat the civilians first. Recall that on deontological views on ethical triage such as Impartiality and TMR, triage ought to respect the right to fair access in a way that promotes the most good. So, in cases such as our current one, where it is fair that either party be treated first, the tie-breaker comes down to which choice would produce the most good. And it seems plausible to claim that treating the civilians would do that even if we assume that medical need and prognosis are the same. This thought could be responsible for this additional fundamental principle of the ICRC:

> Neutrality: In order to continue to enjoy the confidence of all, the [ICRC] may not take sides in hostilities or engage at any time in controversies of a political, racial, religious or ideological nature (ICRC 2015).

Assuming it is a good thing for independent organizations such as the ICRC to have unfettered access to combat zones to provide necessary medical treatment and aid, it seems reasonable to worry that warring parties might obstruct this access, particularly if the ICRC is not seen as neutral. And while treating combatants does not amount to "taking sides," it could certainly be taken as such, especially if they are

---

combatants who follow the rules of war (e.g. targeting only combatants) ought to be discriminated against regarding medical care and aid.

[12] Recall my discussion in 6.1 on this point.

prioritized over one's own civilians. Thus, we might have an additional reason, which could serve as a tiebreaker, to prioritize the treatment of civilians.[13]

8.8 Conclusion

In this essay, I have argued that Impartiality, the moral principle, is false. While medical triage that considers medical need is certainly appropriate, morally speaking, we ought to consider more. According to *Triage by Moral Responsibility*, other things being equal, we ought to prioritize treatment of those who are less morally responsible for being involved in a situation where someone must suffer harm. We ought to allow the more responsible others to suffer the harm. Again, this is not to say that those who ought to suffer the harm deserve to do so. If that were true, then we would have reason to allow them to suffer harm even outside of LRSs. Moral responsibility becomes a criterion for triage precisely because we cannot treat everyone. As with the ethics of defensive harming, this is a matter of distributive justice, or the fair distribution of harm when someone must suffer it.

In the context of the current COVID-19 pandemic, I have suggested that even if it is true that individuals have the moral right to refuse vaccination, it can still be ethically permissible to medically discriminate against them on that basis. I have also suggested that vaccination programs that prioritize high-risk individuals such as smokers over healthier ones are morally defensible even if TMR is true. And this is due mostly to complex questions regarding the *risk* of harm and the challenges for answering them.

Lastly, in the ethics of war, I have suggested that TMR is consistent with the traditional view that we ought to prioritize medical treatment of noncombatants over that of just combatants. Although it is true in some cases that both parties have an equal level of responsibility for being victims in an LRS, it is likely that we have additional reason to prioritize noncombatants, as this will likely cause more good.

References

American Medical Association. 2021a. *Access and Health Equity During a Pandemic*. https://www.ama-assn.org/delivering-care/ethics/access-and-health-equity-during-pandemic. Accessed 12 Aug 2021.

———. 2021b. *Disparities in Health Care*. https://www.ama-assn.org/delivering-care/ethics/disparities-health-care. Accessed 12 Aug 2021.

[13] Another possible moral justification, suggested by an anonymous reviewer, for prioritizing treatment of noncombatants over just combatants is that the latter incur a duty to suffer the harm in an LRS by virtue of occupational choice. Tadros (2020, Ch. 11) presents a very plausible version of such a view, but he ultimately rejects it on grounds of fairness. However, I do not think his arguments are decisive.

Frowe, H. 2015. Reductive individualism and the just war framework. *Legal Philosophy between State and Transnational Seminar Series* 46. Available at http://digitalcommons.osgoode.yorku.ca/transnationalism_series.
International Red Cross and Red Crescent Movement. 2015. *The Fundamental Principles*. http://icrc.org. Accessed 12 Aug 2021.
———. 2021. *Customary IHL Database*. https://ihl-databases.icrc.org/customary-ihl/eng/docs/v1_rul_rule110#. Accessed 12 Aug 2021.
Kagan, S. 1997. *Normative ethics*. New York: Routledge.
McMahan, J. 2005. The basis of moral liability to defensive killing. *Philosophical Issues* 15: 386–405.
———. 2009. *Killing in war*. Oxford: New York.
Quinn. 1989. The doctrine of doing and allowing. *The Philosophical Review* 98 (3): 287–312.
Rachels, J. 1975. Active and passive euthanasia. *New England Journal of Medicine* 292: 78–80.
Tadros, V. 2020. *To do, to die, to reason why: Individual ethics in war*. New York: Oxford.
Thomson, J. 1971. A defense of abortion. *Philosophy and Public Affairs* 1 (1): 47–66.
US Department of Defense. 2017. *Medical Ethics in the Military Health System*. Available at https://health.mil/reference-center/policies/2020/04/22/dha-ipm-18-012. Accessed 12 Aug 2021.
Walzer, M. 1977. *Just and unjust wars*. New York: Basic Books.
Woollard, F. 2015. *Doing & allowing harm*. Oxford: New York.

Lieutenant Colonel Stephen N. Woodside is an Academy Professor of Philosophy at the United States Military Academy, West Point, where he is the Director of the Philosophy Program and teaches various courses in introductory philosophy, ethics, logic, metaphysics, epistemology, and philosophical methods. He completed his PhD in philosophy at Rutgers University in 2016 with a dissertation titled "Liability, Responsibility, and Ineffective Threats." He has presented at various conferences and published in the Journal of Military Ethics and the Notre Dame Philosophical Review on topics related to his continuing research in the ethics of harming and war. Prior to his transition to academia, he served for 17 years as an aviation officer and UH-60 Blackhawk pilot, with an operational deployment to Bosnia-Herzegovina and two combat deployments to Iraq. He currently has over 27 years of active-duty service in the US Army. The views expressed in this paper are the author's. They do not necessarily represent those of the US Department of Defense, the US Army, or the US Military Academy.

Chapter 9
Medical Rules of Eligibility – Can Preferential Medical Treatment Provisions Be Ethically Justified?

Daniel Messelken

Abstract In emergency situations and while medical resources are sufficient, doctors are expected to prioritize and treat patients according to medical criteria only. In MASSCAL situations and when medical resources become insufficient, patient selection and prioritization changes. Rules of triage are applied with the aim of getting the best result possible under the circumstances, e.g., saving the largest number; collective health outweighs individual health. Still, according to the standard ethical principles, non-medical criteria should never influence the doctors' decision of who will be treated. In military contexts, so-called medical rules of eligibility (MROE) provide rules that include non-medical criteria. This chapter aims to analyze if, why, and under what circumstances it may be ethically acceptable to prioritize patients according to non-medical criteria.

Keywords Resource scarcity · Triage · Medical rules of eligibility · Distributive justice · Military medicine · Military trauma care

9.1 Introduction

This chapter discusses ethical aspects of the distribution of scarce resources by military medical services during armed conflicts and other emergencies. More precisely, it analyzes the rules and priorities for access to medical treatment used in many military missions, so-called Medical Rules of Eligibility (MROE), and asks the question if or under what conditions they are morally justifiable. MROE are produced during the planning phase of a military mission and cover the aims and responsibilities of the medical part of the military operation. They set out a

D. Messelken (✉)
Zürich Center for Military Medical Ethics, Center for Ethics, University of Zürich, Zürich, Switzerland
e-mail: messelken@ethik.uzh.ch

S. M. Eagan, D. Messelken (eds.), *Resource Scarcity in Austere Environments*, Military and Humanitarian Health Ethics,
https://doi.org/10.1007/978-3-031-29059-6_9

framework and criteria to decide who will get medical treatment and who will be prioritized if need is. Thus, MROE can be understood as the first step in a triage process. Alternatively, MROE could also be understood as form of resource allocation because the time-pressure associated to triage is not a factor during the elaboration of MROE and because one of the aims of MROE is to avoid the occurrence of triage situations by reducing the number of potential patients. In that understanding, one could also label them as a form of preemptive triage. The latter, however, is achieved by excluding parts of a population from the option to receive medical care in military medical facilities. Consequently, the facilities can then concentrate on treating the prioritized own troops who are always guaranteed unconditional access to treatment.

MROE raise ethical concerns. By (systematically) preferring one patient group over another (that is defined by non-medical criteria) MROE deviate from the standard criteria of medical triage and the standard approach to the allocation of scarce medical resources, which would be based on medical criteria solely. The reasons and arguments for doing so can be different and will be investigated later. The moral question that obviously arises is whether the general, planned, and intentional prioritization of own troops based on non-medical criteria can be justified.

The chapter is split into two parts. It will first explain how MROE are an integral part of the military trauma care system and illustrate this with an example. In the second part, ethical arguments in favor and against MROE will be presented and discussed.

9.1.1 A Cursory Look at IHL – Legal Framework during Armed Conflict

In order to recall the general legal framework under which many military missions (and thus their medical personnel) operate, a brief look at the main lines of international humanitarian law (IHL) may be helpful. The most relevant regulations are to be found in the Geneva Conventions and their Additional Protocols as well as the Customary IHL study of the ICRC. As the latter summarizes the Geneva Conventions as well as later developments, I will only refer to its wording here for the sake of shortness to establish what IHL requires from HCP and what it also allows them when treating the wounded.

Required by IHL. The general rule in customary IHL regarding the medical treatment of wounded and sicked during armed conflict is the following:

> Rule 110. The wounded, sick and shipwrecked must receive, to the fullest extent practicable and with the least possible delay, the medical care and attention required by their condition. No distinction may be made among them founded on any grounds other than medical ones. (Henckaerts et al. 2005, 400)

Two aspects shall be highlighted. First, "with the least possible delay" usually is read to mean that the circumstances must permit HCP to provide treatment without putting their own lives at risk (e.g. during ongoing fighting) but can of course also be read to refer to the availability of resources. Second, the last part of the quoted rule makes it clear that legally no discrimination of patients is allowed if it is not based on medical criteria. The Geneva Conventions are even more explicit and state that "[o]nly urgent medical reasons will authorize priority in the order of treatment to be administered." (International Committee of the Red Cross 2013, I, 12) IHL thus not only allows HCP to triage patients according to medical criteria but even requires them to do so.

Allowed by IHL. Those who are performing a medical role during armed conflict are also explicitly allowed by IHL to follow their ordinary code of medical ethics:

> Rule 26. Punishing a person for performing medical duties compatible with medical ethics or compelling a person engaged in medical activities to perform acts contrary to medical ethics is prohibited. (Henckaerts et al. 2005, 86)

With regard to the question how "medical ethics" as referred to by IHL may be understood in the context of international armed conflict, one can use the *WMA Havana Declaration* (1956, last updated in 2012) or its "update" in the 2015 *Ethical Principles of Health Care in Times of Armed Conflict and Other Emergencies* by the ICRC and other organizations as a point of departure. The most relevant parts with regard to triage of the latter document read:

> In providing the best available care, they shall take into consideration the equitable use of resources. […]
>
> 3. The primary task of health-care personnel is to preserve human physical and mental health and to alleviate suffering. They shall provide the necessary care with humanity, while respecting the dignity of the person concerned, with no discrimination of any kind, whether in times of peace or of armed conflict or other emergencies. […]
>
> 7. In armed conflict or other emergencies, health-care personnel are required to render immediate attention and requisite care to the best of their ability. No distinction is made between patients, except in respect of decisions based upon clinical need and available resources (International Committee of the Red Cross (ICRC), World Medical Association (WMA), International Committee of Military Medicine (ICMM), International Council of Nurses (ICN), und International Pharmaceutical Federation (FIP) (2015).

To summarize, one can thus state that HCP, military or civilian, both during peace and in armed conflicts or other emergencies are (i) allowed or even expected to act according to medical ethics and that (ii) medical ethics remains the same independently of the circumstances. Circumstances obviously play a role during the decision-making process of for example triage, but they shall not alter the underlying ethical principles.

9.2 Medical Rules of Eligibility – Macro-Triage in the Military

In this section, I will introduce to "Medical Rules of Eligibility", a concept commonly used in military medicine/military planning and that serves as form of pre-triage of patient groups to reduce the number of eligible patients and to avoid actual triage scenarios. I will first briefly outline to the extent necessary for an understanding of MROE how military trauma care is organized (9.2.1), and then present two concrete examples of MROE to illustrate how they work in practice and how they may lead to ethical issues (9.2.2).

9.2.1 Military Trauma Care

To understand why MROE are seen by military planners as an essential building block of military trauma care it is helpful to get an insight on how military trauma care is organized. Obviously, no normative conclusions can be drawn out of the following description, but they may help to understand how the problem of MROE and the ethical challenges analyzed later emerge. (For a more detailed account and general assumptions on planning priorities see e.g. Bricknell 2014a; Bricknell 2014b; Bricknell and Gadd 2007.)

When military forces are deployed, they bring their own medical services with them so that they can immediately receive emergency treatment. The mission and tasks of the military medical personnel are planned accordingly:

> Medical support to military operations is tailored to the military population at risk (PAR) and the assessed risks to deployed military troops […]. (Bricknell 2014a, 455)

Nowadays, there is an expectation (and often a legal obligation) that military HCP are able to provide to their own soldiers the same standard as in their home country. Obviously, the local standard of care differs from country to country, and it is effectively much lower in many countries where military forces intervene. The health system in Afghanistan, for example, has a lower standard of care than European health systems. As a result, foreign military medical services sometimes offer the highest standard of medical care in the country they are deployed to, and their abilities can exceed the locally available level of medical care by far.

To meet the aim of providing highest-level medical treatment to the defined group of people it is responsible for, the military trauma care system is organized in a continuum of several layers with very different capacities and only a small portion of them are actually deployed (cf. National Academies of Sciences, Engineering, and Medicine 2016, 100). Treatment starts in the field with the aim of curing as many patients as possible but if need is patients are evacuated within very short delays to receive the required medical care. The approach is summarized in the following statement:

> The philosophy for the treatment of battle casualties used by the British Forces has two principles:
>
> 1. That as many as possible receive treatment and are returned to duty.
> 2. That those who cannot be rendered fit only receive such resuscitation and surgery as is needed to guarantee their safe evacuation to the next echelon of treatment.[1]

[1] https://ihl-in-action.icrc.org/case-study/united-statesunited-kingdom-medical-treatment-and-care-wounded-soldiers-during-persian [last accessed on September 14, 2022].

Consequently, the capacities for high standard treatment can be rather small in mission theatre as they are only planned for the intermediate care and stabilization of wounded soldiers before their (air) evacuation to long-term treatment facilities (usually fixed military hospitals at home or in allied countries). Thus, an own soldier receives emergency treatment in the field hospital and is then evacuated within 24–72 hours after being wounded to receive definitive care "at home". Neither in the planning nor in practice does he occupy the deployed ICU capacities for a long time, and this is why there is no need to deploy large ICU capacities. However, this system can only work if the capacities of high-level initial care in role 1–3 facilities (those who are deployed) are always available and not occupied for longer periods of time by individual patients. Otherwise, situations of a second sort triage would sooner or later become inevitable as the very limited ICU capacities constitute an important bottleneck in the system.

In reverse conclusion, patients would destabilize the system if they were admitted for a long-term ICU treatment to a role 1–3 facility without the perspective of a timely transfer to a larger long-term facility. Long-term treatments are just not planned for in deployed facilities. But local (civilian or military) systems often cannot provide ICU treatment to "their" potential patients at all. Therefore, and to avoid beginning a treatment that cannot be continued adequately in the local system, local patients are usually excluded from treatment in the deployed high-level medical treatment facilities of the intervening military forces at all. Admission to medical treatment in such scenarios does not only depend on the medical condition, but also on non-medical factors such as nationality.

To make this point more tangible, I propose to look at a decision tree/flowchart that visualizes the MROE used at a certain point of time by ISAF in Afghanistan. Its illustration in Fig. 9.1 is mostly a reproduction of a flowchart published by von Uslar (2018, S10), which I have amended to better visualize the two-tier system it results in.[2]

Let us start with the original figure and first ignore the amended elements in blue color on the left. If a wounded local person is brought to the military medical system, the decision about the care she is eligible to receive depends on her function, identity, and type of injury. The options are put in the first line. Embedded interpreters or civilian casualties caused by ISAF would thus be directly considered for treatment, whereas for other groups like national security forces or VIP patients a request would first be directed to the local health care facilities. In any case, all potential local patients are in a second step only admitted to the military medical facility if the MROE state shown in the green square is green (or maybe yellow). This is where the ISAF soldiers as (potential) patients and the blue amendments to the original flow chart come into play. They are prioritized in two ways. First, the "own" wounded and sick (i.e. ISAF soldiers) always have direct access to the military

[2] All amendments are marked by the use of blue color. It is unclear, if the original figure was published by ISAF or whether it was created by von Uslar to illustrate an ISAF document.

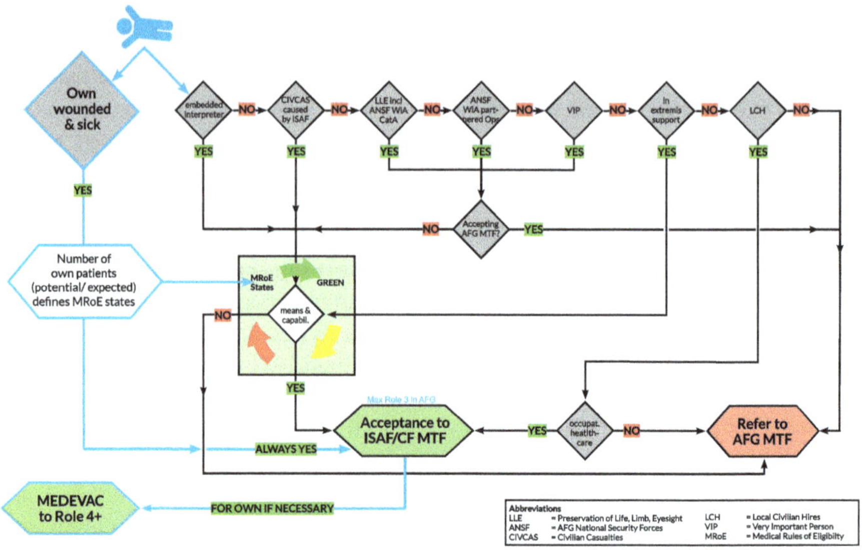

Fig. 9.1 MROE in Afghanistan. (Adapted by D. Messelken from: von Uslar (2018). NB: The blue lines/boxes and their labels are amendments to the original figure)

medical facility and additionally also to MEDEVAC to a higher echelon of care if needed. This is displayed by the blue line on the left side. Secondly, the number of *actual and expected* own patients influences the MROE states relevant for the admission of other patients. In reality, this often means that (ICU) beds are kept empty and resources reserved in order not to risk scarcity in the case that own patients arrive. Obviously, during conflict there is always some risk that own forces are wounded and consequently some medical resources have always to be available as an emergency reserve.

Thus, the military trauma systems used by ISAF and similarly in other missions *de facto* establish different standards of care depending on the group affiliation or group membership of (potential) patients:

> Identity is definitive. There are no restrictions on the care of Coalition forces. For all others, care is contingent upon an imminent threat to life, limb, or eyesight. (Gross 2017, 41)

This raises the ethical question whether group membership should be accepted as a valid argument to justify the prioritization of (or access to) medical treatment. One commonly accepted analogy from civilian life may come to mind: it is often assumed that in extreme situations it would be ethically allowed (or even required) to treat wounded HCP first so that they can join their colleagues in administering health care themselves. The discussion later will show if this analogy is a helpful one or not.

9.2.2 *Example of Applying MROE*

To illustrate the so far rather theoretical description of MROE, I would like to look at a concrete case that describes a situation where civilian access to military health care is evaluated. The following scenario has been reported by a Dutch medical officer and describes a situation that occurred during the ISAF mission in Afghanistan.[3]

Scenario 1

A few days after having set up a role 2 facility with two ICU beds, "NLD paramedics treated civilian AFG patients who had been involved in accidents, all of whom were poly-traumatized. After surgical treatment in the role 2 treatment facility, two patients remained in critical condition requiring intensive care, that they initially continued to receive in the facility. An audit of AFG hospitals in the surrounding area showed that no local facility was able to provide adequate care, and the patients would have died immediately after handover. At the same time, however, military plans for high-intensity operations against insurgents (INS) were well advanced; the NLD commander on site demanded that all the capacities provided be kept ready for the foreseeable occurrence of wounded. To make matters worse, the follow-up supply of medical supplies was slow to get underway and the two AFG intensive care patients were increasingly consuming the supplies of medical oxygen, among other things." (Adapted/translated from von Uslar 2018, S7)

The ethical issues that arise in this scenario are (at least) the following: (i) how long do you continue the emergency medical care in the ICU and (ii) can you move a patient to a facility that you know can only provide insufficient follow-up care to free your capacities for potential future patients? The MROE that (later) were published to provide a basis for decisions in such cases (see Fig. 9.1 with decision tree according to ISAF SOP 01149 above) would only have helped to some extent. They would have allowed the admission of the civilian accident victims for emergency care if at that point of time the available resources had been sufficient (MROE state "green"). It is unclear however, what they would say about continuing the ICU care if the MROE state changed to yellow or red as the commander's intervention in the scenario suggests. The aim of the MROE in ISAF SOP 01149 is to reduce the number of potential (=eligible) patients so that every fully eligible patient (i.e. every coalition soldier) *always* gets a free bed without any delay. In a small medical unit like the one described in the scenario above (ICU capacity = 2) this would usually mean to keep the beds empty rather than to admit a patient with a lower eligibility.

[3] The date of this case is not mentioned in the article, but the mission during which it occurred lasted from 2006 to 2010 (https://en.wikipedia.org/wiki/Task_Force_Urozgan last accessed on 15 September 2022). It thus happened before the ISAF MROE referred to in this paper were enacted but may nevertheless serve to illustrate the ethical issue as similar cases certainly occurred after 2011.

9.3 Can Medical Rules of Eligibility Be Morally Justified?

After having established what MROE are and why they are deemed to be necessary from a practical planning perspective, I will now turn to the questions, if, why, and under what conditions MROE and the preferential medical treatment of only a selected group of people can be ethically justified.

To be sure, the situation that underlies the establishment of the MROE and the following discussion is that treating everybody in medical need would be overwhelming the military medical facilities and exceed its resources by far. Under such circumstances, preference has long been given to own soldiers, even though such an approach *knowingly* collides with the framework of international humanitarian law:

> It must be remembered that the Geneva Convention mandates that enemy casualties be treated in a similar fashion; however, traditionally, U.S. combat casualty care has been directed toward U.S. casualties first, allies second, civilians third, and enemy fourth. (Swan and Swan Jr 1996, 451)

One can of course question the applicability of the Geneva Conventions depending on the nature of conflict and depending on whether the states involved in the conflict have ratified them and the Additional Protocols at all. In this chapter, we are anyway interested in the *moral* perspective on preferential treatment of own soldiers: Can MROE rely on a sound *moral* justification? I will split the discussion into two parts starting in Sect. 9.3.1 with arguments to justify priority for own soldiers from an ethical perspective before turning to counterarguments in Sect. 9.3.2.

9.3.1 Pro – Why Preferential Medical Treatment for Own Soldiers Is Legitimate

Among the arguments most often defended in the literature to ethically justify a preferential treatment of own soldiers over enemy soldiers and other persons in need are military necessity, the fiduciary obligation of HCP to contribute to the overall aim of winning a war, and associative duties towards "brothers in arms". To these arguments, I will add other arguments that have so far not been treated much in the literature but could provide equally valid points.

9.3.1.1 "Military Necessity" or Military Triage Reasoning

The argument most often invoked to explain why military HCP cannot solely or primarily comprehend themselves and act as independent health carers is that of military necessity. At least during conflict, so the argument goes, military HCP are first and foremost soldiers and as such must comply with the exigencies of the military world rather than those of the medical sphere.[4] With regard to triage and MROE, this implies that military criteria dictate the order of treatment instead of medical reasoning. Gross summarizes this approach as follows:

> During war, medical personnel do not treat individual soldiers qua discrete patients but as components of a fighting force: a living, collective entity. To maintain this force medical personnel bear the obligation to salvage soldiers—that is, return as many to duty as quickly as possible. The principle of salvage favors combatants over noncombatants, the less wounded over the severely wounded, and friend over foe. (Gross 2006, 142)

It is however unclear which ethical arguments are behind this beyond a pure utilitarian reasoning with the aim of winning a war (which then must be deemed to be a just war). And even if this aim was accepted, one would still have to show that the selective medical treatment is the only or by far the best option available and that the ultimate aim of winning the war otherwise is in danger. There are certainly situations when this is the case, but as the wording "military necessity" suggests, such situations would rather be exceptional battlefield circumstances that do not endure over a long period of time. In addition, the suffering and loss of life of those not treated adequately on both sides also must go into the utilitarian calculus.

The military necessity argument in my opinion is much less convincing as a general one to be used as a justification during the planning of MROE as this would imply to assume a permanent state of military necessity.[5] With regard to the concrete example in scenario 1, one would have to assume that saving the lives of the potential victims in the planned (=future) military operation are necessary for the military success and that this can only be achieved by discharging the civilian patients who are currently occupying the ICU beds.

As these implied assumptions are not convincing, one has to take into account the possibility that the term "military necessity" is used to cover other arguments (or that they are included in the term). It could for example be used as an umbrella term to refer to internal cohesion, comradery, trust etc. In short then, the aim of MROE justified by military necessity would then consist in supporting the morale of the troop in general rather than to purely look at saving one's own numbers. If this was the case, one should however try to openly argue with these aims rather than rely on "military necessity" and thereby contribute to a questionable use of the term. Further below I will make a proposal to that effect.

---

[4] The ethical challenges of this role conflict are usually discussed under the heading "dual loyalty conflicts" but without a definitive answer to the problem. See for example Allhoff (2008).

[5] As a side note: In IHL, the term "military necessity" has a clearly restricting meaning: attacks etc. can only be allowed if they are necessary and no less harmful alternative exists. In philosophy, on the contrary, necessity is often invoked to justify exceptions that exceed what would be allowed under normal circumstances.

9.3.1.2 Fiduciary Obligation to Win Wars

A second argument goes in a similar direction than the military necessity argument but has an even wider focus. It states that the military as a state-sponsored organization must fulfill the tasks it has been given by the state: usually to win wars (or to resolve a conflict etc.). Adams refers to this obligation as a "fiduciary obligation" (Adams 2008, 224ff.) because the military is entrusted to accomplishing it on behalf of the state. He further states that this task is not given to the fighting components of the military alone, but to the military organization as a whole, which includes the medical services. Adams summarizes his argument as follows:

> …military commanders are entrusted with the duty to win wars. Thus, they possess fiduciary obligations to accomplish this goal without attending to other possible goals they as individual persons might have, unless doing so would enable them to better accomplish their primary obligation, namely winning wars. […] I will suggest that military decisions in general should be governed by an overarching fiduciary obligation to win wars. In particular situations, this may be accomplished by triage models that are egalitarian in focus, but in others commanders must reverse triage procedures to fulfill their obligation and perform the duty with which they have been entrusted. (Adams 2008, 227)

One can criticize the argument for similar reasons than the military necessity argument. There are certainly scenarios in which military medicine can be a decisive factor in winning a war and thus contribute to the fiduciary obligation of the military. But this will only be the case in exceptional situations, and it is again difficult to see how this could help in justifying a *general* approach such as MROE. The example scenarios used by Adams in his article *are* extreme cases and would not challenge the respect for medical ethics and medical triage criteria in more ordinary circumstances, even during conflict.

In addition, it is also relevant how one understands "winning the war" and the fiduciary obligations different people and professions are entrusted with. If one does not reduce "winning the war" to the military defeat of an enemy but considers a long-term perspective of rebuilding peace, this would also open space for arguments not to implicate the medical services too strongly in a purely military logic. A wider understanding of what it means to win a war (in the long term) could allow HCP to continue working according to the civilian logic of medical ethics while still contributing their part to the war effort (cf. Messelken 2017).

9.3.1.3 Associative Duties

A third argument to justify preferential treatment of own soldiers is provided by Gross under the label of "associative duties". According to this argument, "military medical personnel often sense a special obligation to care for compatriots" and justify priority for their comrades because "He's our brother." (Gross 2017, 43) It is certainly the case that the cohesion within military units (or even within military coalitions) is and needs to be very strong, and that values or virtues like comradeship, trust, and the like play an important role for an effective military organization.

It is also clear, that being under enemy fire or similar existential threats foster bonding into groups, and that such an established bond remains strong and influences even others in the unit who have not been in that situation themselves. All this however rather provides psychological explanations why doctors might *want* to help their comrades first, why they *feel* an obligation to do so, but it does not automatically provide them with a *moral justification* to do so.

As an illustration for his argument and the practical relevance of it, Gross relies on the following example:

> One U.S. soldier and one Iraqi Army [allied] soldier present with a gunshot wound to the chest. Both have low oxygen saturations. There is only enough lidocaine for local anesthesia for one patient, and only one chest tube tray. One will get a chest tube with local anesthesia, and the other will get needle decompression and be monitored by the flight medic. (Gross 2017, 43)

This case, however, has a limited value for the discussion and ultimately cannot prove very much. The example is ethically rather unproblematic, as both wounded soldiers get a medical treatment and none of them eventually suffers from a worse long-term harm as a result of not getting any treatment or of not getting a treatment good enough. The second soldier will certainly suffer more immediate pain and discomfort due to the suboptimal medical treatment, but this is a comparably small discrimination. If the argument of "associative duties" is used only to decide in cases like this and as an alternative to throwing a dice, and if the reach of the "associative duties" argument does not go (much) further than this (as Gross (2017, 45) eventually seems to acknowledge) it seems an acceptable argument – as long as one would accept the same argument when made by the other side. Arguing with "associative duties" would then however not be helpful regarding the more complicated example of scenario 1 and not be helpful regarding establishing generally valid MROE that go much further than just *slightly* preferring one patient when medical criteria do not provide a basis for distinction.

However, accepting the argument in some cases instead of relying e.g. on a lottery can prove to set a dangerous example: in the field and with less time and distance, it will be difficult for deployed military HCP to correctly estimate when associative duties can be invoked and when they should rather not rely on the argument. Similarly, potential patients may perceive the action and selection of HCP as biased and inconsistent. Thus, it may be better not to defend and apply the associative duties argument too prominently if one agrees with its limited applicability.

9.3.1.4 Missing Reciprocity: The Other Side Does Not Provide HC or Does Not Respect HCP

Another line of argument (that in some way is based upon a contractarian argument for triage) is based on an assumption of reciprocity as the basis of an obligation to provide medical care. Accordingly, one only owes another group equal medical care if they provide a similar level of care themselves in return, or at least make an effort to do so. When such reciprocity is not returned, there is no such thing as a contract in reality and, consequently, one cannot be bound to it.

In conflict scenarios like ISAF/Afghanistan and when fighting against non-state actors it is indeed often the case that the latter do not provide similar and impartial medical care. This may be due either to a lack of resources and capabilities or due to ideology and consequently lead to different evaluations. The argument and its point might be even more evident if the other side regularly and deliberately attacks medical facilities and thus does not respect them in the first place. This is another sad reality of current conflicts although such attacks are certainly and unfortunately perpetrated by state actors, too (see e.g. UN Human Rights Council 2013). It would seem odd if the same people who attacked medical facilities could later claim a right to receive medical care from these or related facilities based on the argument that medical care should be provided to everybody without distinction. But let us look at both parts of the argument in more detail.

The argument of reciprocal medical care makes a valid point, and a similar reasoning in fact made the signing parties agree to the legal regulations of IHL.[6] Namely, that an acceptance of reciprocal duties to care for all wounded leads to better and more humane results for all. In the reality of contemporary armed conflicts and with regard to our two examples from the ISAF mission (scenario 1 and the MROE in Fig. 9.1) the argument of reciprocity is a bit flawed and thus cannot fully convince. Rather, it puts a spotlight on some contingencies of contemporary conflict theaters, which are the very reason for the establishment of MROE. In most (or even all) military conflicts, the intervening party is the better equipped one and brings the (most of the time much) higher standard of medical care. This is certainly the case for the military missions that Western military forces were involved in recently and for all so-called "humanitarian interventions". Consequently, reciprocity and a comparable level of medical care on all sides is practically impossible and the intervening soldiers in fact would not want to be treated in a facility not of their own side, at least not for more than emergency care.[7] This was different when the Geneva Conventions were agreed upon and when the dominating kind of war was between states and their armed forces, similarly equipped both in weaponry and emergency medical services.

In current conflicts or military interventions, the situation can be further complicated by legal provisions like the prohibition to provide "material support to terrorism," (cf. Eckenwiler et al. 2015; Palama 2022). Such provisions render it even less likely that any current "enemy" would be capable (even theoretically) of building up a health system with a comparable standard of medical care, even if they were willing to invest the necessary financial resources. If the latter was the case, access to

[6] Legal obligations however do persist even if reciprocity is not given. The fact that someone is stealing from me does not give me a right to steal from him in return. The same is true in IHL.

[7] This also may lead to perceiving neutral actors like MSF as belonging to one side: they treat those from the other side, but not ours. The reason that they do not treat military from all sides is however not their taking sides, but the still better care provided by the military medical services of one side.

resources of medical care would still be severely impeded by such laws and make it improbable that the level of medical care could reach a high standard.

This leaves us with the second part of the argument, namely that those who attack HCP cannot claim the same right to treatment as those who respect them. This is a valid argument but mainly concerns a specific subgroup of the population and potential patients for two reasons. First, it can only be applied to combatants or those who perpetrate attacks against medical facilities but not to the civilian population in general. Therefore, it would not be applicable in our scenario 1 as the patients concerned are civilians who do not have an active role in the conflict. The argument could only explain why, e.g., in Afghanistan one might be morally justified not to treat a wounded insurgent who just attacked medical personnel when the medical resources would also be needed to treat an own soldier, even if he was less severely injured. Second even if one accepts the argument that repeated attacks against healthcare facilities make one lose the right to medical care, one must be careful not to judge a whole group of combatants collectively and forego any individual distinctions. At least, if it serves as an argument for MROE that exclude *all* combatants from one side, because *individuals* of their group regularly attack medical facilities. One would have to prove (or legitimately assume) that the attacks on healthcare facilities are systematically used by the other side and agreed to by a large majority of their combatants. Otherwise, one would penalize some individuals for the action of others and hence act morally questionable to say the least.

9.3.1.5 Unjust Combatants Have No Claim to HC

The last part of the previous argument merits some more attention and elaboration in a different variation. It can be aligned to some arguments defended within so-called "Revisionist Just War Theory" (RevJWT) which claim that "unjust combatants" should not be provided medical care. Although this is not the place to dig into the debate on RevJWT, let me briefly explain its most important assumptions that are necessary for our context.

According to RevJWT, we should distinguish between just and unjust combatants, the latter being those who fight for the morally unjust cause or side in a conflict (McMahan 2004). Hence, unjust combatants do not fulfill the requirements of *ius ad bellum* and as a result they do not have a (moral) justification to fight for their cause. Consequently, according to RevJWT "unjust combatants" do not even have a moral right to defend themselves against just combatants.

But does this also mean that they do not have a right to receive medical care once they are wounded and (are forced to) stop fighting? Some authors clearly argue so and justify this with the potential unjustified threat that unjust combatants will constitute once their health has been restored (Frowe 2014, Sect. 8.2; McMahan 2004, 710 f.). If this argument is correct, unjust combatants could be excluded from access to medical care and MROE doing exactly this would only be logical. From my perspective, this conclusion is oversimplified, and several arguments speak against the summary exclusion of *all* combatants of the unjust side from medical care. First, the revisionist argument used as a basis for MROE presupposes that one can judge the whole group of unjust combatants collectively. It ignores, for example, the fact that some individuals may have been forced to fight on the unjust side. Thorough accounts of RevJWT do make these distinctions and would not support summary judgments of all combatants on one side. Second, one must at least distinguish between unjust combatants who can continue to fight and those who will not be able to return to the battlefield anyway. Not treating individuals of the latter group seems inhumane, although it would not be mandatory under all circumstances (e.g. not during ongoing battle). Lastly, Hereth (2021) has convincingly argued that the high risk of unjustified harm being committed by the just side gives good reason to treat unjust combatants as well to meet compensatory responsibilities.[8]

9.3.1.6 MROE Are a Precondition for Military Interventions that, Overall, Still Bring a Better Outcome for the Local Population

Maybe the most promising way to justify MROE vis-à-vis a civilian population that is potentially denied access to medical care by these MROE is by taking the larger context and alternative options into account. Overall, so the (contractualist) argument, the situation of the civilian population is still better if the (humanitarian) military intervention takes place even if they are denied access to the high-level medical care of the intervening military forces. The argument uses consequentialist assumptions to construe consent to a fictitious contract between the intervening military and the civilian population who eventually profit from that intervention. Let me spell out the assumptions in more detail.

The first assumption is that we are dealing with a military intervention that is in the interest of the civilian population (or at least a large majority of it) of the country where the intervention takes place. This is what is usually assumed for humanitarian interventions and in situations where the "responsibility to protect" (R2P) is invoked (International Commission on Intervention and State Sovereignty et al. 2001). Before the intervention, the civilian population is suffering from a regime and this is what the military intervention aims to improve and is a promising means to achieve.

Second, it is assumed that the intervening party does not have an own (dominating) stake in the conflict but intervenes in the interest and on behalf of the local population. The intervening military thus already accepts a risk, which cannot be weighed against an own benefit. To mitigate the risks associated to their military intervention, they bring their medical services with them, and they rely on their high-level care. Now, if the military was not allowed to treat their own wounded

[8] Even if one does not agree with the counterarguments, the RevJWT argument does not apply to MROE excluding "innocent" civilians from medical care. The exclusion from medical care of civilian victims who do not participate in the conflict cannot be made on the same ground than that of unjust combatants (if one accepts the unjust combatant argument at all).

preferentially, the whole mission would not take place, as their soldiers are only willing to take the risk associated to the military intervention if immediate medical care on the level they are used to is guaranteed.

The third assumption and final part of the argument is that without the military intervention the overall situation of the civilian population would be worse, and this would lead to an even less acceptable outcome than accepting different levels of access and different standards of medical care based on group affiliation.

Consequently, the civilian population could (based on a consequentialist reasoning) accept the very restricted access to the intervening party's medical facilities and the establishment of a (temporary) two-tier medical care system as they hope to gain a long-term advantage from the military intervention. To some extent, the argument can be understood as a sharing of the burdens that come with the military intervention, both on the side of the intervening military forces and the population it shall support. The argument does not provide much guidance for individual HCP who find themselves situations like in scenario 1; but it can serve as a moral justification for the establishment of restricted access rules fixed in MROE like the one presented in Fig. 9.1.

This argument overlaps with what often happens in reality when MROE in invited missions are part of a contract/memorandum of understanding (MOU) and thus accepted by the government of the local state on behalf of its population. The analysis above has shown how such a contract can be based on (consequentialist) moral considerations and hence interpreted as a necessary compromise to achieve an overall better outcome.

9.3.1.7 Local Responsibility, Do Not Interfere with Local System

Another line of argument why the MROE are morally justified disputes the responsibility of the military HCP for treating local civilians. More concretely, one would directly refer civilian patients to local health facilities (and the local standard of care) with the argument that the national health care system is responsible for providing medical care to local people. One would in addition argue that it is better to provide support to the local system to create a solution that is sustainable in the long term. In scenario 1 such support could for example consist in sharing medical material (if that is compatible with the resource scarcity described in the scenario).

The argument however has a restricted scope of application and does only work in the context of ordinary medical care that is not related to the military intervention. For example, accidents as in scenario 1 or also chronic diseases. By contrast, the intervening forces would have to take the responsibility to treat cases and compensate additional burdens to the health care system caused by themselves (CIVCAS). Thus, one might discuss whether patients with conditions that are directly conflict-related (such as injuries from crossfire) can be said to only fall under the responsibility of the local health system or if the intervening military forces rather share some responsibility here as well. In fact, when we look at the

MROE as defined in Fig. 9.1, during ISAF the responsibility of the coalition to treat civilian casualties was acknowledged (even though under the condition that resources are not needed by coalition soldiers).

9.3.1.8 Avoid Moral Injury Among HCP

A last argument that we want to look on and that is used to justify MROE and the preferred treatment of own soldiers focuses on the military HCP themselves. It is assumed that if they cannot treat those they are or have become close with first but may be compelled to first treat those that have attacked their comrades or even themselves maybe only minutes ago, this could challenge their moral convictions and lead to moral injury.

However, this argument has also been used to delegitimate MROE providing strict rules for admission and does not allow military HCP to act according to medical ethics. It then reads that if HCP cannot treat those that they want to treat considering medical criteria and medical ethics principles but are compelled to first treat a certain group, this could challenge their moral convictions and lead to moral injury. This would for example be the case if they are not allowed to provide (emergency) care to civilians in need.

9.3.2 Contra – Why Preferential Medical Treatment Is Ethically Problematic

We will now turn to an analysis of the arguments used to morally argue against the moral acceptability of MROE and against the preferential treatment of own soldiers. These counterarguments can be roughly divided into three clusters: general arguments referring to the principles of humanity and non-discrimination, consequentialist arguments about the effective use of resources, and arguments establishing a moral responsibility to provide medical care to patient groups excluded by MROE.

9.3.2.1 MROE Are Against the Principles of Humanity and Non-discrimination

The first and most obvious argument against MROE and the resulting unequal access to medical care is that they contradict the principles of Humanity and non-discrimination, two fundamental principles of both IHL and (medical) ethics. By applying MROE one must accept the outcome of unequal chances of survival in different groups of patients, which are defined on the basis of non-medical criteria. This, however, goes against the principle of Humanity according to which all human beings must be respected equally (Corn 2021). In their famous 1958 Commentary to the Geneva Conventions, the authors (commissioned by the ICRC) go so far as to state that the "obligation to grant protected persons humane treatment is in truth the *leitmotiv* of the four Geneva Conventions." (International Committee of the Red Cross 1958, Art. 27) Therefore, MROE that do not hold up to the idea of humane treatment but rather discriminate individuals because of their group adherence should not only be judged from an ethical point of view, but must also be questioned from an IHL perspective.

The MROE outcome of unequal chances of survival depending on group membership is also contrary to ordinary medical ethics, which only allow for discrimination between patients based on medical criteria. MROE also are in contradiction to the principle of justice, which is among the fundamental ethical principles of medicine and requires a fair distribution of chances and risks (see e.g., Beauchamp and Childress 2019, Chap. 7). It seems very unlikely that we would accept a set of MROE *within* our societies[9] and the fact that they are highly discriminatory is probably a good reason for this. The fact that MROE are only applied during deployments and regarding treatment decisions affecting members of another group of persons should make us skeptical about their moral justifiability. The application of triage rules, on the contrary, is also accepted within societies[10] (and can be justified in medical ethics) and this is due to the fact that they discriminate between individual patients only based on medical criteria and without looking at group criteria. This illustrates the importance that is usually given to non-discrimination and equal respect of persons not only but specifically in medical ethics. And this is an important argument why MROE, other than medical triage, are morally problematic.

9.3.2.2 MROE Lead to an Ineffective Use of Resources – Too Much Reserve Capacity

A second line of argument is a more consequentialist one, referring to the (in)effective use of resources. According to this argument, MROE often lead to a waste of resources or at least a suboptimal use of available medical means. This is because, in contrast to a triage approach, MROE lead to holding back resources for the benefit of potential future patients (of an eligible group) instead of using them for the actual benefit of present patients with an immediate need for medical care but who are part of an ineligible group. If the potential future patients never occur (for example because no eligible soldiers are wounded) then not treating actual patients from an ineligible group is obviously less efficient than treating them. Lives could have been saved or suffering could have been reduced but it did not happen as MROE required keeping reserves.

[9] Remember: MROE are typically used during deployment and to distinguish different levels of access for local patients and deployed personnel.

[10] The recent discussions about potential triage situations in the context of the COVID-19 pandemic has however shown that the acceptance of rules for triage can be difficult within societies.

This argument depends on factors like the predictability of potential patient influx and the actual opportunities to resupply resources within a useful timeframe. It presupposes that material resources can be resupplied and that rather a shortage of medical personnel is the main reason for MROE. In a weaker reading one could thus rather take it as an argument for more discretion at the lower level to decide whom to treat. More concretely, MROE should not be too strict but leave enough room for maneuver to those who apply them, namely the HCP confronted with people asking for medical care in their facility. However, as MROE are also a strategic instrument with the aim of guaranteeing the availability of medical care for a specific population, one can question whether HCP at the lower level are in a position to have the necessary oversight to decide when exactly MROE can be applied less stringently.[11]

Still, MROE almost inevitably will lead to situations in which resources are held back for casualties who never occur and in which, consequently, treating those patients who did show up but were deemed ineligible would have led to a better overall outcome. Such situations can however only be identified with certainty in hindsight.

9.3.2.3 Responsibility at Least with Regard to Caring for Some Patients

A third argument against MROE or at least against the preferential treatment of only own casualties refers to special moral obligations vis-à-vis some other groups of patients, or, at least, some individual patients. Taking the example of the ISAF mission or similar kinds of military intervention, the groups of civilian combat casualties ("collateral damage") and local coalition forces are obvious candidates for such moral obligations, although for different reasons. Regarding wounded civilians, it can be argued that there is a duty to treat them if the harm they suffered is a result of the intervention, a direct result of military force. In the MROE example presented above, they would only be treated if the MROE states are in the green area, that is if there is enough reserve capacity for own military casualties even *after* the civilian casualties have been accepted. This may be justifiable with arguments like the "overall still better"-argument (9.3.1.6 above) on a collective level, but responsibility as a result of harming individuals could be an equally strong or even more convincing argument against MROE. I disagree with the argument of Michael Gross (2021, Ch. 7) that victims of unavoidable collateral harm should rather be compensated financially than by giving them medical care. *Optimaliter*, they would be part of an ordinary medical triage system to access medical care since money cannot by them a comparable chance of medical treatment if the local system is not well established.

[11] In addition, the responsibility of having to decide when to admit additional patient groups may put an unfair burden on the medical personnel.

Similarly, the differentiation made by MROE between external and local coalition forces (e.g., German vs Afghan coalition forces, from a US perspective) is implausible. If they are fighting side by side, it seems unfair to not at least give them the same chance of access to medical care independently of their nationality.

9.3.2.4 Health Care as a Remainder of Peace

The last argument against discriminatory and exclusive MROE presented here is based on an idea I have defended elsewhere in more detail (Messelken 2017; Messelken 2019). This idea in a way summarizes the common ground of the arguments against strict MROE, as it stipulates to practice medicine according to peace standards independently of the context, hence even during conflict. The rationale of the argument is a rather old one (which can be traced way back in traditional Just War doctrine[12]) and requires that one prepares, even during war, for a peaceful aftermath and always keeps this ultimate aim of war in mind. Traditionally, this has led to the interdiction of cruel or "unfair" methods of warfare and methods that would undermine any remaining trust in the enemy and his willingness to eventually find a peace agreement. In my reinterpretation of the idea and its application to the context of providing medical care I suggest that upholding a humanitarian space during conflict can be understood as keeping a bridge to peace intact and alive. Impartial medical care, *as an idea and as a practice*, constitutes a remainder of Peace in the middle of conflict. But to do so, it cannot be bound on conditions defined by the conflict. MROE which remain in the logic of conflict by distinguishing friend and foe clearly follow a different rationale then that of peace and risk that medicine becomes part of the conflict.

Although the peace argument shares some assumptions with the other arguments against MROE it is different insofar as its justification comes from a long-term perspective, which goes far beyond the ongoing conflict. War should not dominate over everything, and a prominent exception could and should be the provision and allocation of medical care.

9.4 Conclusions

This chapter analyzed ethical aspects of the distribution of scarce resources by military medical services during armed conflicts and other emergency situations. The main question of the chapter was whether MROE and the exclusion from access to healthcare for certain groups of patients can be morally justified. Several ethical arguments were reviewed, both in favor and against MROE. As a result, even if there was no definite answer, MROE proved to be difficult to justify ethically from a general point of view, and good arguments are needed to justify MROE in specific

[12] For an overview of the Just War doctrine see for example (Lazar 2017).

contexts. Overall, it seems that MROE may only be ethically justifiable as a form of compromise in situations when they contribute to a better outcome all things considered.

Finally, it seems to me that medicine and the provision of medical care should be regarded from a different perspective and be excluded from the logic of war as far as possible. Such a special consideration of medicine does not seem to convince everybody, and it can of course be asked why medical resources and their distributions should be treated differently at all. It seems at least that complaints are weaker when food, technology, or other advantageous resources are not widely shared or distributed more fairly. Maybe this is because the difference medicine can make in individual cases is just so obvious. But maybe we should also ask whether other life-saving resources should also follow a similar postulate and be distributed more fairly during conflicts to mitigate their consequences.

Acknowledgements Financial support for the work on this chapter was granted by the Centre of Competence for Military and Disaster Medicine of the Swiss Armed Forces.

The author would like to thank his colleagues at the Center for Ethics of Zurich University and the participants of the 9th ICMM Workshop on Military Medical Ethics for their valuable feedback and criticism to earlier versions of this chapter. The opinions defended in this chapter are those of the author.

References

Adams, Marcus P. 2008. Triage priorities and military physicians. In *Physicians at war*, International Library of Ethics, Law, and the New, ed. Fritz Allhoff, vol. 41, 215–236. Dordrecht: Springer Netherlands.

Allhoff, Fritz. 2008. *Physicians at war the dual-loyalties challenge*. Dordrecht: Springer.

Beauchamp, Tom L., and James F. Childress. 2019. *Principles of biomedical ethics*. 8th ed. Oxford: Oxford University Press.

Bricknell, Martin. 2014a. Military medical contribution to indigenous (military and civilian) health sector development (HSD) within security and stabilisation operations. In *Conflict and catastrophe medicine: A practical guide*, ed. James M. Ryan, Adriaan P.C.C. Hopperus Buma, Charles W. Beadling, Aroop Mozumder, David M. Nott, Norman M. Rich, Walter Henny, and David MacGarty, 451–475. London: Springer London.

———. 2014b. Military health services support in conflict. In *Conflict and catastrophe medicine: A practical guide*, ed. James M. Ryan, Adriaan P.C.C. Hopperus Buma, Charles W. Beadling, Aroop Mozumder, David M. Nott, Norman M. Rich, Walter Henny, and David MacGarty, 477–485. London: Springer London.

Bricknell, Martin, and R.D.M. Gadd. 2007. Roles for international military medical services in stability operations (reconstruction and development). *Journal of the Royal Army Medical Corps* 153: 160–164.

Corn, Geoffrey S. 2021. Humanity, principle of. In *Max Planck encyclopedia of public international law*. https://opil.ouplaw.com/view/10.1093/law:epil/9780199231690/law-9780199231690-e1810#. Accessed April 21.

Eckenwiler, L., M. Hunt, A. Ahmad, P. Calain, A. Dawson, R. Goodin, D. Messelken, L. Rubenstein, and V. Wild. 2015. Counterterrorism policies and practices: Health and values at stake. *Bulletin of the World Health Organization* 93: 737–738.

Frowe, Helen. 2014. *Defensive killing*. Oxford: Oxford University Press.

Gross, Michael L. 2006. *Bioethics and armed conflict. Moral dilemmas of medicine and war*. Cambridge: MIT Press.
———. 2017. Saving life, limb, and eyesight: Assessing the medical rules of eligibility during armed conflict. *The American Journal of Bioethics* 17: 40–52.
———. 2021. *Military medical ethics in contemporary armed conflict: Mobilizing medicine in the pursuit of just war*. New York: Oxford University Press.
Henckaerts, Jean-Marie, Louise Doswald-Beck, Carolin Alvermann, and ICRC, eds. 2005. *Customary international humanitarian law*. Cambridge: Cambridge University Press.
Hereth, Blake. 2021. Health justice for unjust combatants. *Journal of Military Ethics* 20: 67–81. https://doi.org/10.1080/15027570.2021.1949782.
International Commission on Intervention and State Sovereignty, Gareth J. Evans, Mohamed Sahnoun, and International Development Research Centre (Canada). 2001. *The responsibility to protect: Report of the international commission on intervention and state sovereignty*. Ottawa: International Development Research Centre.
International Committee of the Red Cross. 1958. *Commentary on the Geneva conventions of 12 august 1949*. Vol. IV. Geneva: ICRC.
———. 2013. *The Geneva conventions of 12 august 1949*. Geneva: ICRC.
International Committee of the Red Cross (ICRC), World Medical Association (WMA), International Committee of Military Medicine (ICMM), International Council of Nurses (ICN), und International Pharmaceutical Federation (FIP). 2015. *Ethical principles of health care in times of armed conflict and other emergencies*. Geneva: ICRC. https://www.icrc.org/en/download/file/9567/ethical_principles_of_health_care.pdf. Accessed 15 March 2023.
Lazar, Seth. 2017. War. In *The Stanford encyclopedia of philosophy*, ed. Edward N. Zalta, Spring 2017. Metaphysics Research Lab, Stanford University.
McMahan, Jeff. 2004. The ethics of killing in war. *Ethics* 114: 693–733.
Messelken, Daniel. 2017. Medical care during war: A remainder and Prospect of peace. In *The nature of peace and the morality of armed conflict*, ed. Florian Demont-Biaggi, 293–321. Palgrave Macmillan.
———. 2019. The "peace role" of healthcare during war: Understanding the importance of medical impartiality. *Journal of the Royal Army Medical Corps* 165: 232–235.
National Academies of Sciences, Engineering, and Medicine. 2016. Overview of contemporary civilian and military trauma systems. In *A national trauma care system: Integrating military and civilian trauma systems to achieve zero preventable deaths after injury*, 73–118. Washington, D.C.: National Academies Press.
Palama, Audrey. 2022. Impact of US anti-terrorism legislation on the obligation of non-state armed groups to provide medical care to the wounded and sick under IHL. In *Health Care in Contexts of risk, uncertainty, and hybridity*, ed. Daniel Messelken and David Winkler, 49–76. Cham: Springer International Publishing.
Swan, Kenneth G., and Kenneth G. Swan Jr. 1996. Triage: the past revisited. In *Military medicine*, vol. 161, 448–452. Oxford University Press.
UN Human Rights Council. 2013. Assault on medical care in Syria.
von Uslar, Rolf. 2018. Wer wird behandelt? Reflexionen zu "Medical rules of eligibility" in Operationen des Internationalen Krisenmanagements. *Wehrmedizinische Monatsschrift* 62: S1–S12.

Daniel Messelken is a research associate at the Center for Ethics at Zurich University and leader of the Zurich Center for Military Medical Ethics (www.cmme.uzh.ch). He also serves as Head Ethics Teacher for the Center of Reference for Education on IHL and Ethics of the International Committee of Military Medicine and is member of the Board of Directors of the International Society for Military Ethics in Europe (EuroISME). Dr. Messelken studied Philosophy and Political Science in Leipzig and Paris (1998–2004) and received his PhD in philosophy from the University of Leipzig in 2010. Besides Military Medical Ethics, his main research fields include Just War Theory, the Morality of Violence, Military Ethics, and Applied Ethics more generally.

Chapter 10
Fairness in Military Care: Might a Hybrid Concept of Equity Be the Answer?

Frederic Gilbert, Ian Stevens, and Samia Hurst

Abstract Applying equity to health care is difficult and it is especially challenging when applied to cases that involve urgent military medicine care under resource scarcity. Part of the difficulty centers on the concept of equity itself. It is not clear what the best concept of equity applicable to medical care would be, or that there should be only one, or the same ones, across all levels of military health care. Despite the fact that equity is a key concern in health care, particularly in the age of the COVID-19 pandemic, it may be that there is no single theory of justice that would be most justified for military physicians to use. This paper examines whether a hybrid position of equity might be both theoretically robust and applicable in practice. After briefly introducing the discussion, we outline four major philosophical definitions of equity – (1) Egalitarianism, (2) Prioritarianism, (3) Desertism, and (4) Sufficientism. We then report empirical findings suggesting that a practice-based hybrid concept of equity is used by physicians within the practice of micro-allocation. Our findings will shed lights on ethical justifications and reasoning which should guide medical rules for military and humanitarian health care providers.

Keywords Desertism · Egalitarianism · Equity · Hybrid views · Prioritarianism · Sufficientism

F. Gilbert (✉)
EthicsLab, Philosophy Program, School of Humanities, University of Tasmania, Hobart, TAS, Australia
e-mail: ims28@nau.edu

I. Stevens
Philosophy & Gender Studies / University of Tasmania, Hobart, TAS, Australia
Department of Neurological Surgery / Oregon Health & Science University, Portland, OR, USA

S. Hurst
Institut Ethique Histoire Humanités, University of Geneva, Geneva, Switzerland
e-mail: Samia.Hurst@unige.ch

S. M. Eagan, D. Messelken (eds.), *Resource Scarcity in Austere Environments*, Military and Humanitarian Health Ethics,
https://doi.org/10.1007/978-3-031-29059-6_10

10.1 Introduction

Equity is a key concern in health care, including military medicine. It is a prerequisite for equality of opportunity in general (Daniels 1985). Equity is one of the building blocks of a risk-pooling health care protocol. For example, during epidemic many ethical issues are linked to how civilian and military personnel fairly address the circumstances -from duties to injections (Messelken 2018; Messelken and Winkler 2017). Equity is important for the moral legitimacy of limit-setting decisions and for their subsequent public acceptance (Hurst et al. 2004). An increasing desire for efficiency in healthcare distribution also leads to a heightened risk that equity could be traded-off (Ubel et al. 2000). Similarly, as highlighted in the beginning of the COVID-19 pandemic, increasing demands for medical resources could lead to substantial issues in fair distribution (Liu et al. 2020). Other contemporary examples include allocation of medical resources for an aging population, the novel, but costly, fields of gene and stem cells therapy, and biomaterial three-dimensional printing-related treatment (3D Bioprinting). Such risks increase the importance of ensuring that the importance of equity is not lost in decision-making processes associated with the allocation of technologies.

Applying equity to health care can be difficult even in ordinary medical cases but is especially challenging when applied to cases that involve urgent military medicine care under resource scarcity linked novel biomedical technologies or when there is a global strain on the healthcare system during a pandemic. Part of the difficulty centers on the concept of equity itself given how this term can be applied differently in varying intra and international locations (Liu et al. 2020). It is not clear what the best concept of equity applicable to medical care would be, or that there should be only one, or the same ones, across all levels of a health care, including military protocol, at which resource allocation occurs.

The idea that the hybrid or pluralist view of equity is needed for resource allocation is a point which has been repeatedly made in the health ethics literature (Cookson and Dolan 2000; Persad et al. 2009). This particularly rings true in contemporary bioethical literature because of the historic COVID-19 pandemic causing medical supply shortages around the world (Verkerk and Lindemann 2011; Liu et al. 2020). It has also been pointed out, however, that, "combining principles into systems increases complexity and controversy" (Persad et al. 2009, 426). This paper examines whether hybrid positions of equity might nevertheless be both theoretically robust and applicable given the widespread utilization of other multidimensional ethical frameworks, such as Principlism (Beauchamp and Childress 2019). We examine this question in four steps. After briefly introducing the discussion, we outline four major philosophical definitions of equity – (1) Egalitarianism, (2) Prioritarianism, (3) Desertism, and (4) Sufficientism – and then examine each of these theories through an ethical scenario involving the example of 3D bioprinting-related treatment. Not only do we argue that a hybrid view is appropriated, but we also show that such a view is already commonly utilised by physicians in their everyday decision-making (McCullough 2020). To this end, we report and present empirical findings that suggest that this hybrid approach is used by physicians within the practice of micro-allocation. We finally examine how robust such hybrid views are by exploring their theoretical weakness and strengths. If overall this argument is successful, pragmatic views of equity would be better supported and prioritized for scare resource allocation concerns.

10.2 Equity in Bedside Resource Allocation

When it comes to making medical decisions concerning resource allocation, most physicians are likely not well versed in the philosophical debates of equity and how it might apply in such cases. With or without the knowledge of these explicit philosophical arguments, however, the social currency arising from various accounts of distributive justice implicitly informs their decisions in their respective practices, both for the better and for the worse (Supady et al. 2021; Verkerk and Lindemann 2011). Ideally, the application of theories of equity in the relevant decisions made by physicians should function in such a way as to reduce arbitrariness – after all, rationing medical resources by clinical judgment is often controversial because it may be susceptible to subjectivity and bias (Hurst and Danis 2007). More recently some authors have advised against the physicians caring for those patients with COVID-19 from deciding these ethically complex questions around resource allocation (Satomi et al. 2020) while others have encouraged it (McCullough 2020; Supady et al. 2021). It is therefore important to know which concepts of equity doctors ought to use, to help them to practice micro-allocation in a fairer manner. Currently, the normative philosophical literature on distributive justice revolves around two questions: first, 'what benefit are we distributing?' that is, the metric debate; and second, 'how should such benefits be distributed?', that is, the non-metric debate. The first question deals, for instance, with notions like resources (Rawls 1971), welfare (Arneson 1989), and capabilities (Sen 1995). This paper will focus on the second question.

10.3 Traditional Concepts Employed in the Debate

In this section, the philosophical literature pertaining to egalitarianism, prioritarianism, desertism, and sufficientism, is briefly discussed by applying these different theories to a particular ethical scenario surrounding the implementation of novel biotechnologies.[1] There is no special reason why these positions are analysed in this

[1] In light of the global pandemic and the resulting plethora of bioethical literature surrounding the equitable distribution of ICU beds, hospital ventilators, and COVID-19 vaccines, this chapter will focus on what the authors suspect may be more prominent future ethical concerns. However, the contemporary literature surrounding fairness is still included in the chapter's argumentative framework.

Table 10.1 Four traditional views conceived separately

Equity concept	Egalitarianism	Prioritarianism	Sufficientism	Desertism
Position's content	Equality is the best distribution.	Priority is given to worse off.	Everyone must have enough.	*Each person should receive treatment corresponding to her accomplishment.*
Distribution	Global distribution.	Singular distribution.	Singular distribution.	Singular distribution.
Reasons	Inequality is intrinsically bad.	To be badly off increases priority.	Focus on people's level of insufficient Well-being.	*Deservingness* presupposes that individuals' virtue is interpersonally measurable.

specific order. Since an exhaustive discussion is not feasible here for each position, this paper provides a cursory interpretation of each position and examines how each plays out in decisions made in the contexts of limited resources in the health care system. The main features of each position are outlined in Table 10.1.

To illustrate this ethical scenario, consider the case of a physician needing to make a choice about two young patients, both of whom require a lifesaving treatment. One of the two is in a significantly better-off situation (medically speaking), however both require an urgent life saving medical treatment that depends on biomaterials custom-fabricated and tailored-made through 3D bioprinting, which raise ethical issues (Gilbert et al. 2017, 2018; Wallace et al. 2014). The two young patients need a 3D printed medical sleeve to be implanted into them due to a life-threatening bronchial tube birth defect. Traditional treatments cannot save either of the young patients' lives. The treatment requires that, prior to being implanted, a CT scan of the patient's airways need to be taken. The data from the scan will be used to 3D print a biocompatible integrated replica cast that will be fully resorbed during the following years (Zopf et al. 2013). Now, however, imagine that due to an insufficiency in resources that only one of the two candidates can be saved, but not both. Indeed, the only 3D printable biomaterial is indivisible: half would save neither patient.

In such a situation, how does the physician reach a decision? On what philosophical grounds can or should she base her decision? How does any of the previously mentioned understandings of equity help us resolve this dilemma? The following section presents four positions of equity, analyses them with respect to the moral conundrum, and discusses some difficulties regarding what elements should be used by doctor when making this resource allocation decision.

10.3.1 Egalitarianism

An egalitarian acts as someone who cares about equality, and considers that "it is in itself bad if some people are worse off than others".[2] Temkin (Temkin 2003) writes that egalitarianism does not only concern itself with equality but must seek equality per se, that is, as a "non-instrumental" aim (Fleurbaey 2015).

The positive effect of this view for physicians is that all other things being equal, no patient must be treated differently. This outcome indicates that inequality is bad because it has bad effects for some patients. By eluding inequality, physicians, somehow, elude bad effects and produce better circumstances. This often occurs when levels of better off patients reach levels of worse off patients. This approach can be thought off as a counterbalance to utilitarian principles that may future perpetuate societal inequalities (Supady et al. 2021).

Nevertheless, eliminating inequality might be problematic to translate into the dilemma of two candidates for 3D bioprinted lifesaving treatment as one of the two candidates can only be saved. In this specific example, inequality cannot be reducing by diminishing the level of the better off patient until he reaches the level of worse off. Should this be the case, the doctor would need to show it is morally good that, between two candidates for the lifesaving 3D printing that is indivisible, the patient in the better situation should be rendered worse off because the other could not be offered better treatment.

This difficulty has often been described as the *levelling-down objection. This objection is* often the biggest challenge for egalitarists to overcome when facing micro-allocation. The challenge is to demonstrate that how fighting inequality by diminishing patient's treatment in order to treat worst off contains something intrinsically good (Broome 1991). The difficulty for physicians is to show that the levelling-down objection does not necessarily imply that inequality ended when treatment of those who are better off are limited or deprived, rendering them as worse off as well for guaranteeing a sense of equity. Limiting access to treatment is surely problematic in medicine, with emerging technologies such a 3D bioprinted-related treatments that are based on personalised medicine. For example, it would be hard for a physician to defend that a better off patient should be deprived the possibility of using her own stem cells for a printed replacement bone. In view of the facts this would seem to be unfair.

However, the egalitarian view of equity has a substantial role to play in allocating novel health resources on macro level, especially given they pose a significant expense to healthcare systems. Although, it should be noted that this does not simply entail access to curative treatments, but, as can be seen with the recent pandemic, entail quality of care improvements like palliative care (Ballantyne et al. 2020). Similarly in age of blossoming biotechnologies, promotion of egalitarian values can theoretically support the benefit of a universal access to personalised medicine healthcare.

---

[2] More precisely, this is what Parfit calls the Telic egalitarianism. See (Parfit 1997, 204).

10.3.2 Prioritarianism

Prioritarianism maintains the view that "benefiting people matters more the worse off the people are" (Clayton and Williams 2000). In some respect, when benefiting others, the worst-off patient is to be given absolute priority over the better off (Crisp 2003, 752).

Here, the positive outcome for a physician is that each patient's point of view must be taken separately to achieve a decision that would be significant enough to be acceptable to each patient, regardless of equality. In this regard, equality has no value. Broome supports this point of view with his principle of personal good. He writes, "if one distribution gives some persons more wellbeing than another distribution does, and if it gives no person less well-being than the other does, then it is better than the other" (Broome 2015).[3] Prioritarianism stands in contrast to egalitarianism or methods that allocate resources randomly (e.g. a 'lottery' system) due to its emphasis of the unique qualities of an individual (Leclerc et al. 2020).

In the context of the above dilemma concerning the 3D printed sleeve the physician must choose which of the two patients to save. Prioritarianism applies weights to benefits (outcomes). Using this theory, should the immediate benefit be higher in saving the medically worse off patient then the physician could decide to treat him without knowing whether the long-term benefits are optimal. However, such a decision would pose a problem well described by Daniels (Daniels 1994) namely, how much priority should be given? In theory, this principle could dictate that doctors should give *complete* priority to those whose health state are the worst, even if they fail to help them in any significant way, and sacrifice in their attempt the opportunity to help many others.

That been said, acknowledging the importance of treating a badly off patient has the medical advantage of determining who should benefit more from the treatment rapidly. Time being crucial at the micro-allocation level, prioritarianism his certainly an important approach serving decision of allocation limited resources. However, this is not a principle that must be strictly adhered to for beneficial results, as some authors have emphasized that prioritarianism does not supersede the clinical experience garnered by attending physicians (Leclerc et al. 2020).

10.3.3 Desertism

In this view, to deserve to be given some specified form of treatment a patient must possess some characteristics, or have done something, which justifies the treatment. Feinberg refers to such characteristics as "desert bases" (Feinberg 2000). The fact

[3] This definition by Broome is quite similar to Rawls's principle of difference. Fleurbaey also calls it the Pareto principle. For Brown, the major distinction between Absolute Prioritarianism and Rawl's difference principle is that the latter is defined in space of "primary goods" rather than that of wellbeing. See Brown footnote 11 in (Brown 2005).

that a person should *receive treatment according to her* "desert bases" brings a distribution depending on how someone's deserts compare to others patient.

The positive outcome of such position is that, if *deservingness is intimately associated with equity,* it allows physicians to only treat patients who deserve it. For instance, in the above dilemma, suppose that there are some justifiable claims of personal desert for one of the two patients. Let us pretend that the urgency of the worse off patient, that is, his need to receive a 3D printed medical sleeve bronchial implant, is the result of a risky illegal endeavour. Let us suppose he tried to escape police after been caught in an illegal act. His bronchial condition deteriorated drastically following exceptional cardio-vascular efforts to run away. The better off patient requires a treatment due to a natural and progressive deterioration of a pre-existing bronchial abnormality. On a deservingness approach, the physician's dilemma becomes a choice between saving a patient who might morally deserve a treatment more and saving one who might deserve it less.

However, it is important to remember that the role of desert in the determination of fair distributions must be kept in proportion in a world where individuals are born unequal. Even a perfect meritocracy cannot satisfy this basic request: to deserve a treatment, you need to have gained your desert from an equal start and through an equal trajectory. Perhaps in the above dilemma, some will think that the young rescuer deserves a treatment because he deserves those assets that made the rescue possible. Because he is more worthy in this sense, he deserves treatment. However, it might be argued that his character depends in large part upon external social and familial circumstances for which he cannot claims credit (e.g., his parents encouraged him to do sport since being young) (Rawls 1971).[4] We tend to overestimate the role of personal merit in health outcomes, and even when this role exists serious problems remain in using deprioritization as a form of sanction (Clavien and Hurst 2020). According to this, the idea to base a treatment on desert seems to be difficult to justify.

With respect to micro-allocation, then, the deservingness argument may be both too complex and inappropriate from an ethical and procedural standpoint. This is especially so in cases where physicians need to make an ethical decision based on nearly anything considered desirable provided only by a patient owning some worthy arbitrary characteristics or some actions (Young 1992).

10.3.4 Sufficientism

Sufficientism represents the following view: "what is important from the point of view of morality is not that everyone should have the same but that each should have enough" (Frankfurt 1987, 21). Instead of pursuing equality, Frankfurt suggests employing, "available resources in such a way that as many people as possible have

[4] For similar argument, see these two works (Nagel 1973; Wasserstrom 1976).

enough or, in other words, to maximise the incidence of sufficiency" (Frankfurt 1987, 31).

The outcome for physicians consists in the assertion that, "if every patient had enough treatment, it would be of no moral consequence whether some had more than others" (Frankfurt 1987, 21). Sufficientism, however, has often been criticised for allowing arbitrary thresholds. For example, when we try to convert these sufficientist values into our ethical dilemma above, we see it is crucial that a decision be made as to *where* an adequate level of care should be drawn. When thinking of 3D bioprinting-related treatment, a sufficientist physicians may have a challenge to draw a conclusive line for *how much treatment is enough* for one of the two patients. A quick answer might be saving one life is enough saving, but it would not clearly indicate to the doctor who should receive the treatment.

The arbitrariness of the threshold is a difficult problem to resolve. However, sufficientism has the advantage to guide physicians to concern themselves with providing a baseline level of care instead of wondering whether all patients should receive the same care. The minimum requirement is that the doctor administer enough care such that the patients are comfortable, and all easily available avenues be taken to relieve their condition. But still, without indication of the minimum requirement, sufficientism will need to be enhanced by some other theory of justice to resolve dilemmas of this type (Crisp 2003, 758).

10.4 On Which Philosophical Ground Physicians Base Their Decision?

As they are described in the philosophical literature, it would be naïve to expect these four positions of equity to be widely embraced by physicians. Physicians are already expected to have a basic understanding of medical ethics, and, unless a focus of a medical ethics course, theories of justice are likely not to be thoroughly taught to all medical students. Previous research has shown that physicians are able to point to instances of inequity in their working environment (Hurst et al. 2007, 137) and that they use a variety of concepts for equity in their day-to-day practice of resource allocation. In that respect, when making resource allocation decision, physicians use some sort of theory of justice. Although the US is known it's use of principle, this can further vary based on a physician's geographic location (Verkerk and Lindemann 2011). The question is what theory or theories of justice are they using?

Based on an interview study conducted of thirty-one Swiss primary care physicians, which is more extensively described by one of us elsewhere (self-identifying reference), it was found physicians used a hybrid view of justice when making resource allocation decisions. The result of this study suggested that physicians were amalgamating key ideas from these four different conceptions of justice to address specific clinical problems that required a resource allocation decision. In describing situations where they make a resource allocation decision in various clinical settings, and how they weighed concerns for equity in reaching these decisions, physicians expressed views of equity that include mixing different approaches. One interesting aspect is that physicians specifically reported the importance to them *as such* of using multiple criteria, rather than a single criteria of justice, in reaching these decisions. These finding supported physicians taking a non-dogmatic understanding of equity. Their view was to balance understandings of equity, what we call a *hybrid position* of equity. In the words of an outpatient internist:

> *(…) little by little, with every day that passes, we realize that we can do the same things, or take the same problem, in different ways and this is where probably we get closer to, to a notion of equity, meaning that we can take several criteria into account at the same time (…)*

Briefly, but illustratively, in intensive care:

> *(…) It's [equity] a little bit a mixture of all this (…)*

A hybrid conception of equity may therefore be viewed as a check list for assessing intuitive conclusions and allows physicians to briefly question each position when making rationing decisions at a micro level. Physicians seemed in fact to be combining several things to make what appeared to them the most pragmatic solution. First, they combined consideration regarding what they were allocating. This is a concern more related to the metric debate mentioned above. The resources they had to share included treatments, but also results, human contact, access to care, and even support for other social goods, such as academic achievement in the case of children needing surgery to improve their hearing (self-identifying reference).

Secondly, they combined criteria on which to base their decision. The criteria they mentioned could be categorized as physical (disease severity, urgency, prognosis, and the like), relational expression of autonomy (such as differences in patients' insistence for receiving treatment), understanding of their prognosis in relation to other patients, the emotional impact they had on their care team and the socioeconomic contexts (such as insurance coverage, but also social standing). The combination of these criteria described by physicians was two-tiered: on the one hand they reported efforts to be as inclusive as possible with physical criteria, and on the other they reported efforts *not* to take relational and social criteria into account. This was reported as difficult, not so much because of pressure on them to do so, but rather because these criteria tended to be unconsciously accepted and socially ingrained principles that were already accepted outside of medicine. This was in tension with the physicians' professional commitment to leave such natural impulses to one side. The criteria they *did* use were a mix of objective and subjective criteria based on scientific data and their clinical knowledge, but also on the general impression a patient gave them. They repeatedly reported changing their minds after seeing a patient personally, for example because their judgment that prognosis was poor based on the patient's file was no longer so clear after seeing the patient's energy and 'will to fight'. Indeed, using a comprehensive mix of objective and subjective concerns was explicitly named as a one of the ways in which physicians attempted to make clinical resource allocation fairer.

Thirdly, physicians reported that of the several views of equity outlined above, they applied none of them systematically and none of them in a unitary fashion. Physicians often described how priority was given to the greatest medical need or to the greatest urgency. They rejected giving priority to those whose personal characteristics did not place them amongst the medically worse off. For example, those who asked for treatment louder, or who were socially closer to the physicians, did not receive greater weight in their assessments. Physicians sometimes voiced egalitarian considerations in the choice of criteria. For example, giving the same care to patients who are relevantly similar based on their physiology and disease, but dissimilar based on their insurance coverage, was viewed as an intrinsically good thing even though it was a goal considered difficult to achieve.

Physicians also spent a long time in the interviews explaining how patients with private insurance coverage, who may have more comfortable rooms, were not medically better off than the other patients. This can be viewed as an egalitarian concern, since it is exceedingly rare that these patients' care could affect that of other patients in any way, even if they *were* provided with better care. Sufficientist views were also expressed as a kind of safeguard. For example, when patients were uninsured or could not pay, there was a minimum physicians felt they *must* do, and for which they make huge efforts to mobilize resources. Although doing *only* that will leave them with frustration, it was considered significantly and identifiably less bad than in cases where they *cannot even* reach this threshold. Even desertist views were expressed. For example, some interventions were the object of negotiations where part of the agreement was that a patient must contribute to their own care by smoking cessation to allow treatment options to not appear futile. Desertist views were also expressed with perplexity in the case of a young foreign woman who seemed to have come to the country already ill, and who needed a liver transplant. In that case the notion that she needed to have contributed to the local social group before 'deserving' a transplant was voiced, but ultimately set aside.

As outlined above, equity as a comprehensive concern for objective and subjective scrutiny is not found in the philosophical normative literature. Consequently, a hybrid view of equity is not addressed in the literature per se. However, physicians seem to mix not only different valid claims for resources, but also claims that may seem less valid than others. Due to uncertainty as to which concept of fairness is best, results of this study suggest that physicians are applying different (though often poorly articulated) theories of fairness simultaneously, in very much the same way as they might prescribe different antibiotics when the identity of an infectious agent is unknown. Despite the imperfections of this approach, it does constitute a kind of hybrid theory of equity that is commonly applied in medical practice. Similarly, the apparent deployment of this hybrid theory supports previous arguments contenting the inevitability of combining theories of justice when trying to encompass all morally relevant considerations (Hurst 2011). Although this practice-based hybrid position of equity is poorly delineated in relation to each theory of equity, it may also be useful for those, such as ethics consultants, who might be called upon to help physicians improve fairness in their allocation decisions (Hurst et al. 2004).

10.5 Exploring Hybrid Equity: Are Hybrid Concepts Robust Enough?

What we have learned from above is that the four philosophical approaches may better serve physicians in their resource allocation decisions when used in a combined manner. Even though the philosophical literature emphasizes the incompatibility of these different positions, what is true of their ethical justifications may not hold for their application. In practice, why should each conception of equity be conceived separately? In fact, there might be good theoretical reason not to work exclusively and separately with each of the four positions of equity. For instance, Fleurbaey claims that a "prioritarian will always find some egalitarian view which reaches the same practical conclusions about all possible cases, although possibly for different reasons" (Fleurbaey 2015). In that sense, he believes that the thrust of the theoretical distinction of equity is about foundations, not about practical applications. Fleurbaey's point demonstrated that a single practical conclusion can emerge from multiple theories. Although there is little explicit philosophical literature about a hybrid view, (Crisp 2003, 758), (Brown 2005) and (Casal 2007) and others[5,6] have made a major step in this debate. An overview of possible categories of hybrid combinations of equity is sketched out in Fig. 10.1.

Outside the recent surge of fair allocation concerns with the COVID-19 pandemic, hybrid views were not typically discussed in the philosophical debate regarding equity (Ballantyne et al. 2020; Leclerc et al. 2020; Liu et al. 2020; McCullough 2020; Satomi et al. 2020; Supady et al. 2021). The last few decades of philosophical literature concerning distributive justice have involved authors taking stands on one position or another, rather than trying to find compatibility between them. Most of the arguments presented were initially made to reinforce a particular point of view by criticising others views and, in turn, responding to received criticism. This pattern is a traditional *exclusivist* approach. Against such a background, opponents of hybrid views might argue that, by defending a *non-exclusivist* argument, hybridians are simply *pointing out* problems in other theories and that this - in the end - does not resolve the apparent incompatibility of the different positions.

The pointing out argument, which can be made by opponents of a hybrid theory, highlights the fact that hybrid views may not really solve anything because they do not deal with each position's core problems. In that context, hybrid theorists may be accused of superficiality: hybrid views simply do not go deep enough as they avoid each view's core problem and do not adequately address each views' fundamental

[5] Also, the possible compatibility between prioritarianism and egalitarianism has been discussed by few philosophers, see (Peterson and Hansson 2005; Temkin 2003; Rabinowicz 2003).

[6] Although it is not purely a hybrid, Huseby's sufficientism has some analogies with Casal's suggestion of sufficiency-constrained prioritarianism. Huseby view defends that below the threshold, a form of prioritarian reasoning might be adopted to specify how distribution should be conducted. See (Huseby 2010).

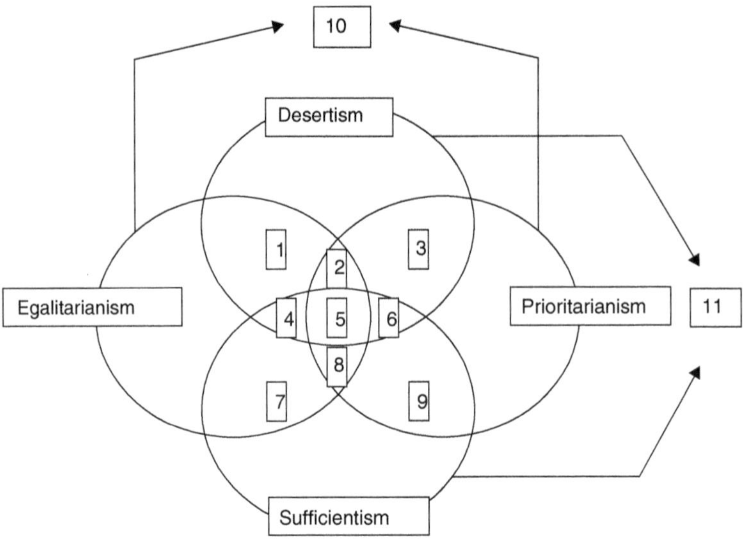

Fig. 10.1 Emergence of hybrid views of equity

difficulties. Thus, they are not theoretically robust enough and they cannot be justifiably recommended to physicians.

An opponent of a hybrid theory would suppose that a physician in our 3D bioprinting-related treatment scenario cannot base her decision on merit because she cannot know the deservedness of a patient; therefore, she should prefer egalitarianism to avoid unfairness and support equal treatment. Since egalitarianism recommends not treating a patient if there is inequality, she should therefore give priority to the worst off. But prioritarianism forces her to treat the worst off who is going to die anyway even with or without the treatment, instead of treating a less worse off patient and therefore the physician should use sufficientism by introducing a threshold of treatment. However, no one knows how to determine a threshold and therefore she should base their decision on something else. This line of argument can go on without end by constantly pointing at other positions.

The pointing out argument, however, highlights the fact that each of the four positions taken separately provides an insufficient answer. The pointing argument can be turned against the opponent of a hybrid theory. The above objection indicates just how each theory is underdetermined as a single viable solution. Hybridians can focus on what sort of restrictions could help physicians' choice to act according to conceptions of justice, instead of on what sort of restriction underdetermined their decision of treatment (which isolates each position with their core problems) (Beauchamp and Childress 2019). For example, if there are not enough biomaterials to 3D print lifesaving implants for all patients, physicians do not need to wait until every patient might have the same treatment, or to give priority to the worst off, or find a sufficient threshold. Rather than attempting to reach the fairest possible choice, they may use the criteria provided by each single approach to avoid what would constitute *unfair* choices on each of them. In this sense at least, combining theories would have an added normative value. This would not be a question of greater philosophical consistency: exclusivist positions will still be superior in that respect. As a guide to the best ethical choice among possible alternatives, however, hybridization could indeed prove to be the better guide. In combining views of equity across different sorts of practical choices under circumstances of constraint, rather than resolving each position's problems, hybrid views could bring robust outcomes.

The pointing out argument also raises the possibility for each of the four positions to provide an overdetermined answer. Suppose there is a particular case where all independent theories answer with the same solution. In such a case, so would the hybrid view approach. In such cases, however, the very confluence of the different views' conclusions bolsters the proposed solution, and this is a point in favour of a hybrid view as it allows such situations to be identified more systematically. Suppose there is a different overdetermined case, but here all independent theories yield different conclusions. Such examples, however, only strengthen the case for a hybrid view since it can provide a framework to compare the adequateness of every single solution before deciding on the more appropriate – or the least inappropriate – solution. In both these cases of over determination, a hybrid theory increases the scope of the theoretical background and with it the robustness of the decision. Therefore, the overall effect resulting from the use of the hybrid position appear to be solid enough to defeat the pointing argument. Each 'ingredient' of a hybrid view can act as a side-constraint to all the others, restricting the number of acceptable options to the sort of decisions where the practical conclusions of different theories are closest to one another.

Further critiques of a hybrid theory comment that an aim of distributive justice is to identify the single correct position that can optimally help physicians. Perhaps that single concept is among the positions currently proposed and physicians need to determine which concept is the right one. However, hybridians take a different stand for the time-limited practicability and for the robustness of the view. Even if a physician personally believes that there is a single correct concept of distributive justice that is yet to be determined, the pragmatic benefits of hybrid model make the most of the several legitimate candidates. An alternative response to the exclusive position of the four traditional views would be that, while neither physicians nor philosophers may yet have found the correct concept of equity, and even though both need to wait for further progress in philosophy, the hybridian nevertheless provides a temporary way of addressing these medical dilemmas that require a decision be made. Therefore, since the hybridian at least considers all 'candidate' principles seriously, it may represent an improvement over single theory considerations, and may be a sufficiently acceptable approach.

Hybrid views may not help to abolish all core problems associated with each position separately understood, but they can diminish many of them. The question about the hybrid non-exclusivist is whether these apparently different positions are all reducible to one super view that can avoid each position's core problem, or whether we should think that there really are several distinct views, which need to be combined to avoid all core problems in each position.

10.6 Conclusion

Even though arguments employed by each philosophical position on equity look incompatible, and seem to not fit together, a hybrid view may be a robust and applicable alternative in the allocation of health care resources. By using a combination of distributive positions, hybridisation does not lose robustness: on the contrary, it increases its solidity by supplementing weaknesses with side-constraints from other positions. This does not mean that traditional views should be excluded for reasons of complexity, or for yielding results that are difficult to accept in practice. Indeed, a hybrid view helps to understand that physicians should not reject traditional views: it might simply help to understand each position more fully by applying it in combination with other views. The importance of using multiple criteria lifts equity into a flexible but solid practical tool.

A philosophical debate still surrounds hybrid fairness, including the difficulty associated with *how* different views should be combined in practice (Cookson and Dolan 2000; Hurst and Danis 2007; Kerstein and Bognar 2010; Nord 2010). Despite this uncertainty, it may be justified to consider hybrid fairness as many applications of basic knowledge to medicine have been in the past: an imperfect but defensible solution, likely to yield a better result than the status quo, and which it may be good to apply while waiting for better knowledge. Such has been the case over the course of the COVID-19 pandemic as theories of allocation are applied in a setting of scientific uncertainty (Leclerc et al. 2020) and where it may be the case for various novel technologies (Wallace et al. 2014; Viana et al. 2017; Kagan et al. 2022; Harris et al. 2022; Gilbert and Lancelot 2021; Harris and Gilbert 2021; Gilbert et al. 2019a, b). One limit of applying a hybrid view of fairness to clinical care in the current context is that using a hybrid approach requires great clarity. This should give us the impulse to train physicians better in regard to concepts of fairness that are relevant to their practice since, however poorly they articulate it, they are already utilizing a form of hybrid concept in their practices (McCullough 2020).

References

Arneson, Richard J. 1989. Equality and equal opportunity for welfare. *Philosophical Studies* 56 (1): 77–93. https://doi.org/10.1007/BF00646210.

Ballantyne, Angela, Wendy A. Rogers, Vikki Entwistle, and Cindy Towns. 2020. Revisiting the equity debate in COVID-19: ICU is no panacea. *Journal of Medical Ethics* 46 (10): 641–645. https://doi.org/10.1136/medethics-2020-106460.

Beauchamp, Tom L., and James F. Childress. 2019. *Principles of biomedical ethics*. 8th ed. New York: Oxford University Press.

Broome, John. 1991. *Weighing goods: Equality*. Wiley-Blackwell: Uncertainty and Time.
———. 2015. Equality versus priority: A useful distinction. *Economics and Philosophy* 31 (2): 219–228. https://doi.org/10.1017/S0266267115000097.
Brown, Campbell. 2005. Priority or Sufficiency...Or Both? *Economics & Philosophy* 21 (2): 199–220. https://doi.org/10.1017/S0266267105000568.
Casal, Paula. 2007. Why sufficiency is not enough. *Ethics* 117 (2): 296–326. https://doi.org/10.1086/510692.
Clavien, Christine, and Samia Hurst. 2020. The undeserving sick? An evaluation of patients' responsibility for their health condition. *Cambridge Quarterly of Healthcare Ethics* 29 (2): 175–191. https://doi.org/10.1017/S0963180119000975.
Clayton, Matthew, and Andrew Williams. 2000. The ideal of equality. Macmillan.
Cookson, Richard, and Paul Dolan. 2000. Principles of justice in health care rationing. *Journal of Medical Ethics* 26 (5): 323–329. https://doi.org/10.1136/jme.26.5.323.
Crisp, Roger. 2003. Equality, priority, and compassion. *Ethics* 113 (4): 745–763. https://doi.org/10.1086/373954.
Daniels, Norman. 1985. *Just health care. Studies in philosophy and health policy*. Cambridge: Cambridge University Press. https://doi.org/10.1017/CBO9780511624971.
———. 1994. Four unsolved rationing problems a challenge. *The Hastings Center Report* 24 (4): 27–29. https://doi.org/10.2307/3562841.
Feinberg, Joel. 2000. Justice and Personal Desert. In *Rights and reason: Essays in honor of Carl Wellman*, Law and Philosophy Library, ed. Marilyn Friedman, Larry May, Kate Parsons, and Jennifer Stiff, 221–250. Dordrecht: Springer Netherlands. https://doi.org/10.1007/978-94-015-9403-5_13.
Fleurbaey, Marc. 2015. Equality versus priority how relevant is the distinction? *Economics and Philosophy* 31 (July). https://doi.org/10.1017/S0266267115000085.
Frankfurt, Harry. 1987. Equality as a moral ideal. *Ethics* 98 (1): 21–43. https://doi.org/10.1086/292913.
Gilbert, Frederic, and Mathilde Lancelot. 2021. Incoming ethical issues for deep Brian stimulation: When long term treatment lead to a "new form of disease". *Journal of Medical Ethics.* 47: 20–25. https://doi.org/10.1136/medethics-2019-106052.
Gilbert, Frederic, John Noel Viaña, Cathal O'Connell, and Susan Dodds. 2017. Enthusiastic portrayal of 3D bioprinting in the media: Ethical side effects. *Bioethics* 32 (June). https://doi.org/10.1111/bioe.12414.
Gilbert, Frederic, Cathal D. O'Connell, Tajanka Mladenovska, and Susan Dodds. 2018. Print me an organ? Ethical and regulatory issues emerging from 3D bioprinting in medicine. *Science and Engineering Ethics* 24 (1). https://doi.org/10.1007/s11948-017-9874-6.
Gilbert, Frederic, Mark Cook, Terence O'Brien, and Judy Illes. 2019a. Embodiment and estrangement: Results from a first-in-human "Intelligent Brain Computer Interface" trial. *Science and Engineering Ethics* 25 (1): 83–96. https://doi.org/10.1007/s11948-017-0001-5.
Gilbert, F., C. Pham, J.N.M. Viana, and W. Gillam. 2019b. Increasing brain-computer interfaces media depictions: Pressing ethical concerns. *Brain-Computer Interfaces.* 6 (3): 49–70. https://doi.org/10.1080/2326263X.2019.1655837.
Harris, Alexandre, and Frederic Gilbert. 2021. Restoring vision using optogenetics without being blind to the risks. *Graefe's Archive for Clinical and Experimental Ophtalmology.* https://doi.org/10.1007/s00417-021-05477-6.
Harris, Alexandre, Mary Walker, Frederic Gilbert, and Patrick McGiven. 2022. Investigating the feasibility and ethical implications of phenotypic screening using stem cell derived tissue models to detect and manage disease. *Stem Cells Reports* 7: 1–10. https://doi.org/10.1016/j.stemcr.2022.04.002.
Hurst, Samia A. 2011. A step toward pluralist fairness. *The American Journal of Bioethics: AJOB* 11 (12): 46–47. https://doi.org/10.1080/15265161.2011.613527.
Hurst, Samia A., and Marion Danis. 2007. A framework for rationing by clinical judgment. *Kennedy Institute of Ethics Journal* 17 (3): 247–266. https://doi.org/10.1353/ken.2007.0021.

Hurst, Samia A., J. Russell Teagarden, Elizabeth Garrett, and Ezekiel J. Emanuel. 2004. Conserving scarce resources: Willingness of health insurance enrollees to choose cheaper options. *The Journal of Law, Medicine & Ethics* 32 (3): 496–499. https://doi.org/10.1111/j.1748-720X.2004.tb00162.x.

Hurst, Samia A., Reidun Forde, Stella Reiter-Theil, Anne-Marie Slowther, Arnaud Perrier, Renzo Pegoraro, and Marion Danis. 2007. Physicians' views on resource availability and equity in four European health care systems. *BMC Health Services Research* 7 (1): 137. https://doi.org/10.1186/1472-6963-7-137.

Huseby, Robert. 2010. Sufficiency: Restated and defended. *Journal of Political Philosophy* 18 (2): 178–197. https://doi.org/10.1111/j.1467-9760.2009.00338.x.

Kagan, Brett, Daniela Duc, Ian Stevens, et al. 2022. Neurons embodied in virtual world: Evidence for organoid ethics? *American Journal of Bioethics Neuroscience* 13 (2): 114–117. https://doi.org/10.1080/21507740.2022.2048731.

Kerstein, Samuel J., and Greg Bognar. 2010. Complete lives in the balance. *The American Journal of Bioethics* 10 (4): 37–45. https://doi.org/10.1080/15265160903581718.

Leclerc, Thomas, Nicolas Donat, Alexis Donat, Pierre Pasquier, Nicolas Libert, Elodie Schaeffer, Erwan D'Aranda, et al. 2020. Prioritisation of ICU treatments for critically ill patients in a COVID-19 pandemic with scarce resources. *Anaesthesia, Critical Care & Pain Medicine* 39 (3): 333–339. https://doi.org/10.1016/j.accpm.2020.05.008.

Liu, Yangzi, Sanjana Salwi, and Brian C. Drolet. 2020. Multivalue ethical framework for fair global allocation of a COVID-19 vaccine. *Journal of Medical Ethics* 46 (8): 499–501. https://doi.org/10.1136/medethics-2020-106516.

McCullough, Laurence B. 2020. In response to COVID-19 pandemic physicians already know what to do. *The American Journal of Bioethics* 20 (7): 9–12. https://doi.org/10.1080/15265161.2020.1754100.

Messelken, Daniel. 2018. On the duty to care during epidemics. In *Ethical challenges for military health care personnel : Dealing with epidemics*, ed. Daniel Messelken and David T. Winkler, 144–163. London, U.K.: Routledge.

Messelken, Daniel, and David Winkler, eds. 2017. *Ethical challenges for military health care personnel: Dealing with epidemics*. 1st ed. Routledge.

Nagel, Thomas. 1973. Equal treatment and compensatory discrimination. *Philosophy and Public Affairs* 2 (4): 348–363.

Nord, Erik. 2010. Balancing relevant criteria in allocating scarce life-saving interventions. *The American Journal of Bioethics: AJOB* 10 (4): 56–58. https://doi.org/10.1080/15265161003633045.

Parfit, Derek. 1997. Equality and priority. *Ratio* 10 (3): 202–221. https://doi.org/10.1111/1467-9329.00041.

Persad, Govind, Alan Wertheimer, and Ezekiel J. Emanuel. 2009. Principles for allocation of scarce medical interventions. *Lancet (London, England)* 373 (9661): 423–431. https://doi.org/10.1016/S0140-6736(09)60137-9.

Peterson, M., and S. Hansson. 2005. *Equality and priority*. https://doi.org/10.1017/S0953820805001664.

Rabinowicz, Wlodek. 2003. The size of inequality and its badness some reflections around Temkin's inequality. *Theoria* 69 (1–2): 60–84. https://doi.org/10.1111/j.1755-2567.2003.tb00754.x.

Rawls, John. 1971. *A theory of justice*. Harvard University Press. https://www.jstor.org/stable/j.ctvjf9z6v.

Satomi, Erika, Polianna Mara Rodrigues de Souza, Beatriz da Costa Thomé, Claudio Reingenheim, Eduardo Werebe, Eduardo Juan Troster, Farah Christina de La Cruz Scarin, et al. 2020. Fair allocation of scarce medical resources during COVID-19 pandemic: Ethical considerations. *Einstein (São Paulo)* 18. https://doi.org/10.31744/einstein_journal/2020ae5775.

Sen, Amartya. 1995. *Inequality Reexamined. Inequality reexamined*. Oxford University Press. https://oxford.universitypressscholarship.com/view/10.1093/0198289286.001.0001/acprof-9780198289289.

Supady, Alexander, J. Randall Curtis, Darryl Abrams, Roberto Lorusso, Thomas Bein, Joachim Boldt, Crystal E. Brown, Daniel Duerschmied, Victoria Metaxa, and Daniel Brodie. 2021. Allocating scarce intensive care resources during the COVID-19 pandemic: Practical challenges to theoretical frameworks. *The Lancet Respiratory Medicine* 9 (4): 430–434. https://doi.org/10.1016/S2213-2600(20)30580-4.

Temkin, Larry S. 2003. Measuring Inequality's badness: Does size matter? If so, how, if not, what does? *Theoria* 69 (1–2): 85–108. https://doi.org/10.1111/j.1755-2567.2003.tb00755.x.

Ubel, Peter A., Jonathan Baron, Beth Nash, and David A. Asch. 2000. Are preferences for equity over efficiency in health care allocation 'all or nothing'? *Medical Care* 38 (4): 366–373.

Verkerk, Marian A., and Hilde Lindemann. 2011. Theoretical resources for a globalised bioethics. *Journal of Medical Ethics* 37 (2): 92–96. https://doi.org/10.1136/jme.2010.036830.

Viana, J.N.M., M.A. Bittlinger, and F. Gilbert. 2017. Ethical considerations for deep brain stimulation trials in patients with early-onset Alzheimer's disease. *Journal of Alzheimer's Disease*. https://doi.org/10.3233/JAD-161073.

Wallace, G. G., R. C. Cornock, C. D. O'Connell, S. Bernie, S. M. Dodds, and F. Gilbert. 2014. "3D BioPrinting: Printing Parts for Bodies." http://ecite.utas.edu.au/95655.

Wasserstrom, Richard. 1976. The university and the case for preferential treatment. *American Philosophical Quarterly* 13 (2): 165–170.

Young, R. 1992. Egalitarianism and Personal Desert. *Ethics*. https://doi.org/10.1086/293399.

Zopf, David A., Scott J. Hollister, Marc E. Nelson, Richard G. Ohye, and Glenn E. Green. 2013. Bioresorbable airway splint created with a three-dimensional printer. *The New England Journal of Medicine* 368 (21): 2043–2045. https://doi.org/10.1056/NEJMc1206319.

Frederic Gilbert is a Senior Lecturer in Ethics, affiliated with the Ethics, Policy & Public Engagement program of the ARC Australian Centre of Excellence for Electromaterials Science (ACES), located at University of Tasmania, Australia and concomitantly an Ethics Consultant for the Centre for Neurotechnology at the University of Washington, in Seattle, USA. He has published over 70 publications in neuroethics, bioethics, and applied ethics.

Ian Stevens is currently a Master of Research Candidate at the University of Tasmania where he is investigating the global regulatory policies of closed-loop deep brain stimulation. He completed his undergraduate education at Northern Arizona University, in Flagstaff, USA where he graduated with a B.A./B.S. in Philosophy/Biomedical Sciences. Outside of his graduate research, Ian is also serving as a member of the International Neuroethics Society's Student/Postdoc Committee.

Samia Hurst is professor of biomedical ethics and chairs the Institute for Ethics, History and the Department of Community Health and Medicine at the University of Geneva. Her research focuses on fairness in clinical practice and the protection of vulnerable persons, and she is the author of the definition of vulnerability integrated into the Declaration of Helsinki and the CIOMS international ethical guidelines on health-related research involving humans. She is a member of the Senate at the Swiss Academy of Medical Sciences, of the Swiss National Advisory Commission on Biomedical Ethics, and vice-chair of the Swiss COVID19 Science Task Force.

Chapter 11
Cicero and the Problem of Triage: Why There Is No Moral Algorithm in Distributing Scarce Resources

Bernhard Koch

Abstract This chapter recalls an ancient model of applied ethics that distinguishes between four different moral roles and can thereby help clarify the structure of the institutional setup, as well as individual action under conditions of resource scarcity. Even if material questions remain open, there is an independent gain in the structural analysis, from which especially the importance of judgment (*phronesis*) emerges strengthened. In this sense, it also represents a defense of the practice of ethical reflection through case discussions.

Keywords Moral roles · Resource scarcity · Triage · COVID-19 · Military medical ethics · Medical emergency

11.1 Introduction

The Corona pandemic has brought the problem of medical resource allocation under conditions of scarcity to the attention of a broad public more clearly than ever before. Military medical ethics and (medical) emergency ethics have been dealing with the issue intensively for much longer and have had to deal with it. Triage criteria catalogs are constantly controversial, but what seems uncontroversial is that medical personnel may or even must use such catalogs as a guide, and that medical personnel cannot set themselves up as moral judges of guilt and merit with patients.

B. Koch (✉)
Institute for Theology and Peace, Hamburg, Germany

Department of Philosophy, Goethe University, Frankfurt, Germany

Department of Systematic Theology, Albert Ludwig University, Freiburg, Germany
e-mail: Koch@ithf.de

S. M. Eagan, D. Messelken (eds.), *Resource Scarcity in Austere Environments*, Military and Humanitarian Health Ethics,
https://doi.org/10.1007/978-3-031-29059-6_11

Nor will I argue below that physicians and nurses should constantly pronounce any moral judgments on their patients, and therefore I do think they need such catalogs of criteria in the distribution of scarce resources, but I will show that in moral evaluation there is nevertheless something like a "bottom line" for any distributional question. In this sense, it is not the possibility of morally based distribution that requires special justification (or is unjustified), but conversely, it must be asked why we have permission at all to deviate from a purely morally based distribution in the direction of a catalog that focuses on extra-moral[1] circumstances, such as medical neediness. In sum, this contribution is intended to show that we should not expect any immediate practical instructions for medical action from ethical reflection, but that we must always bring judgment to bear as a situational "competence."

11.2 Moral Inequality

11.2.1 A Simple/Simplified Case

Let us take a very contrived and, in this sense, quite unrealistic case description as a starting point: A local, habitual speeder, who is just on his way home from a drinking spree at five o'clock in the morning, collides, due to his recklessness, with the vehicle of a female doctor who has set out so early to help out at the Corona Vaccination Center. The accident has dire consequences: Both the doctor and the speeder are seriously injured. For analytical purposes here, we must further assume that the doctor and the speeder are injured in the same way, i.e., have the same survival probabilities, and that the arriving emergency medical personnel, due to unfortunate circumstances, have with them resources for the sufficient, survival-saving care of *only one* injured person. We blank out everything else for the sake of analysis ("ceteris paribus"). It seems to me that our moral intuitions are quite clear: The doctor who is innocent of the accident should receive these resources, not the reckless speeder. To put it more defensively: We would not blame the emergency physician for allocating the medical resources to the female doctor who had the accident, even though in doing so she, the emergency physician, did not make the medical circumstances the decisive criterion for the allocation. There were no medical circumstances here that could have determined the distribution; the medical circumstances were the same for both accident victims.

[1] I here rely on the traditional (e.g. Aristotelian) distinction of three sorts of goods: goods of the soul, goods of the body, and external goods. Classical ethical theory holds that there is a hierarchy: Goods of the soul (moral goodness, virtues) are proper moral goods and are in this respect more important than goods of the body (e.g. health, strength) or external good (e.g. wealth, honor).

11.2.2 Revisionist Just War Theory

Perhaps, then, in such a stereotyped case,[2] we can agree on the application of the moral criterion, namely, that the person who caused the accident should not be placed in the same position as the innocent victim.[3] But how does it look if there are slight medical differences after all. Let's assume, then, that the notorious speeding driver is injured a little more easily, his chances of survival are not good, but they are somewhat better than those of the doctor who got into this situation completely innocently. Would we be willing to set off the "moral bonus" against the "medical malus" here? Before we go into this question, we should consider the even more decidedly pro-moral position of saying that if I have to allocate a scarce resource between a person who bears no responsibility for the misfortune that makes him in need of that resource and a person who bears that very responsibility for the misfortune, then the innocent person should be preferred over the responsible person in any case-no matter what other extra-moral considerations might be in favor of the person who caused the accident.

We would thus be making a similar consideration here to that made by proponents of so-called "Revisionist Just War Theory" (RevJWT)[4] when they distinguish in armed conflict between persons who are "liable" because they bear responsibility for the violent event and those persons who are without responsibility for the threats that have arisen. Revisionist Just War Theory "holds that we can judge the morality of war by thinking about what individuals are permitted to do to each other."[5] There is no principled distinction between "peacetime ethics" and "wartime ethics". Only the applications change.[6]

For Jeff McMahan[7] and the other proponents of this normative approach, most civilians are in this sense "innocents" against whom violence may not be directed.

---

[2] In an ICMM course on Military Medical Ethics in Times of Armed Conflict (MME) (held in Spiez, Switzerland), one participant brought up an example that is perhaps even more emotionally drastic: Let's imagine a woman is raped but defends herself with the result that both the rapist and the woman are seriously injured and in need of urgent medical attention. Who should a doctor give it to first - given scarce resources? It seems that moral intuition is clear here too.

[3] The thesis that moral responsibility for the situation is the first factor in any distribution under resource scarcity is not based on metaphysical assumptions but could be founded on a hypothetical consensus under "a veil of ignorance" just as well.

[4] There are different kinds of "Revisionist Just War Theories": All of them agree about the asymmetry of unjust combatants and just combatants. But the basis of the distinction can be seen differently, e.g. moral responsibility or culpability or causal nexus. The theory is called "revisionist" because it revises the standard approach of International Humanitarian Law and Michael Walzer's "Just and Unjust Wars" (5th ed., New York 2015).

[5] Helen Frowe: Defensive Killing, Oxford 2014, 123.

[6] In this sense RevJWT is fully consistent with General Principle 1 of the widely endorsed *Ethical Principles of Health Care in Times of Armed Conflict and Other Emergencies* which says: "Ethical principles of health care do not change in times of armed conflict and other emergencies and are the same as the ethical principles of health care in times of peace."

[7] Cf. Jeff McMahan: Killing in War, Oxford: Oxford University Press 2009.

Of the combatants, only those who belong to the aggressive party to the conflict ("unjust combatants") – not those who are legitimately defending themselves and other innocents ("just combatants") – are the ones who may be violently attacked at all because they are liable. The deliberative starting point here is a situation of legitimate self-defense: If an unjust aggressor attacks an innocent victim, the victim may defend herself/himself with violence, but this does not give the aggressor the right to use violence in return. Responsibility for a threat establishes a normative asymmetry between a liable person (or several persons) who must bear the burdens of the situation and an innocent person (or several persons) whose rights remain unaffected. It is not the legal categories of "combatant" and "civilian" based on extra-moral characteristics ("uniform", member of state armed forces) that play the decisive role in the moral question of who is liable to be attacked in a war, but the moral characteristics of moral responsibility. Liability is based not only on intentional threatening, but on responsibility for a threat, which can also result from a person acting negligently or recklessly.

Transferred to the accident situation, it could be said: The person who caused the accident does not have to have intended the damage to the accident victim, but he is nonetheless responsible for this damage. Therefore, she/he must bear the burdens of the situation – which includes the scarcity of the healing resource – and forgo this resource in favor of the innocent victim.

11.2.3 Moral Blackmail

Now, one might object that in real situations it is much harder to say who bears moral responsibility for a threat than our stereotypical case description suggests.[8] This could be taken as an argument in favor of solving the problem of distribution by lot.[9] However, the same objection is made against Revisionist Just War Theory by saying: In real conflicts it is simply not possible to decide which responsibility

[8] Imagine a car driver with poor eyesight: For sure he is not responsible for his poor eyesight (on the contrary: he is to be regretted), but if he nonetheless drives his moral responsibility for any accident based on his poor eyesight is amplified. But many factors of this kind may not be discernible, especially not in the urgency situation itself. But the same thing applies with triage criteria like "remaining life years" (cf. the famous SIAARTI-Recommendations on triage (March 2020) in *Recenti Progressi in Medicina* 111/4 (2020), 207–211; https://www.recentiprogressi.it/archivio/3347/articoli/33183/ [15.03.2023]): A base jumper may be young but none the less may the number of his remaining life years be reduced. The number of "remaining life years" under ordinary conditions *as a base jumper* may be smaller than that of a bookkeeper the same age.

[9] There are some advantages of a decision by sortition, e.g. somebody could argue that human beings in general lack moral responsibility of the relevant kind, because they lack moral agency of the relevant kind. – There is a prominent argument that flipping a coin or throwing dices grants the same dignity to every person involved, but that would be a dignity which leaves moral features of a person out of consideration. Moreover decision-taking by lot is only fair if for all people concerned the same thing is at stake. This again brings back considerations of remaining life years etc.

for the unjust threat specific persons bear. Therefore most "revisionists" do, after all, concede the legal, i.e. extra-moral, distinction between civilians and combatants as a guide to moral judgment. But it is only a guide, a clue, an indicator, which, however, can also go astray.

If one were to rely only on the extra-moral factors, one would get into the peculiar situation where someone could by acting intentionally improve his position in the distributional sequence when resources are scarce. Let's take the initially plausible assumption that prioritization should occur according to need. The neediest person receives first. This would mean that a needy person who sees through his situation and that of his fellow competitors for the scarce resource can improve his position by willfully causing himself additional harm. In this case, he would be worse off due to the additional harm, but would be compensated by moving up in the distribution hierarchy. This could be significant in life-and-death situations.

Or take the equally extra-moral criteriology of "first come, first served". It seems extraordinarily counterintuitive to give a scarce resource to a person who comes in first, knowing fully well that he or she has grabbed the palm by illegitimate means. If two COVID-19 patients need to be urgently transported from location A to the nearest hospital in location B at the same time (e.g. for artificial respiration), but the driver of one patient deliberately knocks the other patient's vehicle off the highway (perhaps at that patient's directive), so that one patient arrives at the hospital first and can take what may be the last breathing apparatus.[10] Yet, intuitively, we would take this to be morally problematic or even impermissible. Thus, even on the question of morally just distribution of scarce resources, a moral judgment must be based on moral criteria (such as moral responsibility, perhaps guilt, perhaps moral liability), but not on extra-moral criteria such as strength or speed, kinship or number of allies, and the like.[11]

[10] It is assumed that the second patient arrives a little bit later. If reckless patient 1 knocked patient 2 fully out there would be no distribution problem anymore, and patient 1 would get the respirator. The deliberation is about resource distribution, not about punishment.

[11] A brief comment on the criterion of remaining years of life: On the one hand, the quantity of years alone cannot be a suitable criterion, since the quality of the years also plays a role. Additional years of physical suffering are probably to be evaluated differently than years of fresh falling in love etc. Quality-of-life assessments are problematic because it is unclear what standard they can be based on in the first place. Michael Quante distinguishes the "naturalistic standard", the "social standard", the "sub-subjective-rational standard" and the "personal standard". (Michael Quante: Menschenwürde und personale Autonomie. Demokratische Werte im Kontext der Lebenswissenschaften, Hamburg: Meiner, [2]2014, 33 ff.). With Bernard Williams (Moral Luck in: B. W.: Moral Luck, Cambridge: Cambridge UP, 20–39) one can still add the temporal factor: The quality of the years depends retrospectively on the value convictions to which one adheres at the time of the evaluation of the past life. Thus, in retrospect, one may regret years of life that were perceived as trouble-free. The assumption "Old people are rich in years of life and privileged, young people are poor in years of life and severely underprivileged, should they die" (Mannino op. cit., p. 46; 54) is only true from a certain value standpoint, which is taken at a certain point in time. Someone may feel that they have lived "bad" years, but that "good" ones are now coming. Another expects only "bad" years after the "good" ones. Again, it would seem that a life is good if it is a morally good life.

11.2.4 Consequentialism

We can now return to the question posed at the beginning of Sect. 11.2.2 about offsetting the moral and extra-moral factors. Suppose, after all, that the speeder is less injured than the accident victim, and medical expertise must come to the judgment that the scarce resource, if used on the innocent accident victim, would be more or less wasted. If the resource is given to the victim, in all likelihood, both parties involved in the accident would not survive: The victim because of the severe and unrelievable injuries, the speeder because he is deprived of the medical resource. In such a case, wouldn't we advocate that the person who caused the accident should get the resource? He should be able to survive. – In a certain respect, this seems to be in conflict with our thesis, stated above, that it is the moral factors that matter: Life is, after all, itself an extra-moral good; why does it now beat the moral aspects?

That life is an extra-moral good does not mean that it is morally inconsequential. As a prerequisite for human freedom and human action, it has special moral weight. It is just that in this consideration we change the direction of view from a merit-based and in this sense deontological consideration to a teleological or consequentialist one. In our initial case (11.2.1), when we still assumed that the medical circumstances of the speeding driver and the accident victim are quite the same, we could have argued from a teleological perspective for the distribution of the resource to the accident victim – not because she is an innocent victim, but because she herself is a doctor and is also committed to the common good (e.g., at the vaccination center). This commitment has inherent moral value as an expression of good character – which leads to good actions.

Now, if the life of the accident victim cannot realistically be saved – seen from a medical perspective –, the distribution of the resource to the victim cannot be justified from a teleological-ethical perspective. However, this does not necessarily mean that therefore the distribution to the person who caused the accident is correct from a teleological perspective. Consequentialist ethical judgment (as in utilitarianism, for example) always requires the assumption of a series of predictions about how the future course might turn out. This also requires the assumption of probabilities for the possible courses, which cannot be easily verified empirically. Every factual course of events presents itself *ex post* as the one that had to occur, so to speak, if one chooses option A at time t. What would have occurred if option B had been chosen can only be postulated (contrafactually) in retrospect, but no longer be ascertained. This epistemic uncertainty is one of the great difficulties of consequentialist conceptions of morality. We cannot say whether the speeding driver, if he cures his accident injuries, will later again massively endanger people on the road, or whether, just the opposite, he will experience a kind of remorse and purification that leads him to drive more carefully. Perhaps, due to feelings of guilt on his part, he might also increase social involvement or participate in accident prevention. As long as one cannot say with great certainty that the speeder will use his survival for renewed bad deeds, e.g., a crime, there is nothing against his survival from a consequentialist point of view, but everything rather in favor of it.[12]

11.2.5 Against Punishment

Thus, if the accident victim's chances of survival cannot be significantly improved by allocating the scarce resource, so that one must assume that the resource would be wasted if allocated to the innocent victim, the triage problem essentially dissolves. We are left with a resource and a person in need of it (namely, the person who caused the accident). Since we cannot identify any *a priori* consequentialist reasons against giving the resource to the accident causer (which would be tantamount to letting him die or killing him by omission), the only question that remains is whether there are deontological reasons against giving it. These would have to be reasons that have as their basis the bad action of the speeding driver that has already taken place and thus justify a punitive sanction. Here, then, the wide field of the problematic legitimation of punishments as a whole opens up: First of all, there is the question of the authority of punishment, which certainly does not lie with the emergency physicians who are called to the scene of an accident, but there is also the problem of the justification of punishment as a whole: Is punishment based on retribution or on a consideration of prevention? If the latter: Is the speeder to be educated as an individual so that he will refrain from speeding in the future (a consideration that does not apply if a vital resource is not provided), or is the general public to be educated on the example of this speeder by showing them what sanction speeders can expect ("social prevention")? We can abbreviate these considerations here by pointing out that we fundamentally reject death or physical mutilation and injury as a sanction for bad behavior. Punishment, then, cannot justify not giving the medical resource to the person who caused the accident.

---

[12] In fact, however, the problem of teleologically justified triage criteria goes even further, as can be well illustrated by the popular model of basing triage criteria on the medical likelihood of success. In Germany, this model has met with great opposition from disability associations (cf. i. a. Annette Dufner: Welche Leben soll man retten? Eine Ethik für medizinische Hilfskonflikte, Berlin: Suhrkamp 2021, 229 ff.; Adriano Mannino: Wen rette ich – und wenn ja, wie viele? Über Triage und Verteilungsgerechtigkeit, Stuttgart: Reclam, 2021, 43–45): If the "prospect of medical success" were to become a decisive criterion, then disabled people would be structurally disadvantaged because their prospects of success might be somewhat limited due to their disability. But this would mean that disabled people are doubly disadvantaged: First, by the disability itself, and second, by being downgraded in prioritization in obtaining scarce resources. Indeed, one could argue that non-disabled people have already had a relatively better life compared to disabled people and should therefore take a back seat in the distribution. On the other hand, however, giving preference to disabled people would mean that the (scarce) resources would not be fully exploited, i.e. wasted, so to speak. It seems to me that this consideration shows that all attempts to base triage criteria only on extra-moral considerations (and both disability and medical chance of success are extra-moral considerations) lead to paradoxical situations.

11.3 Cicero's Concept of Different "Personae"

11.3.1 Perspective Agency

What does this mean for our initial problem of triage? The primacy of the moral consideration with regard to responsibility and liability has not been called into question by the concession of a consequentialist consideration in favor of saving lives and by the rejection of punitive reasoning.[13] Nevertheless, the considerations made so far are one-sided in one important respect: They attempt to establish, as it were, an 'objective moral picture of the situation' and do not take sufficient account of the situation of the agent who is faced with the question of distribution. In our case, this is the emergency physician who is called to the scene of an accident. As Thomas Nagel, for example, has repeatedly tried to show, the subjective perspective of the person acting does not simply merge into an objective view abstracting from the subjective viewpoint. The emergency physician does not take a standpoint "from nowhere", but she stands concretely in a situation that is not abstracted but real, on which she looks from her very specific point of view. It is true that we associate moral action with the claim that we detach ourselves from our merely situational constitutions and adopt a viewpoint that focuses on general characteristics ("moral point of view"), because such characteristics can be communicated. Acting ethically does not mean to take into account only one's subjective and concrete needs or aspirations, but to find general criteria from which the scope and limits of the pursuit of subjective aspirations can be intersubjectively justified. But on the other hand, we do not take the standpoint of an "archangel" (R. M. Hare) from which we can gain a total overview of the situation, so that we could conceive of ourselves only as a variable Y, now subject to an imperative to act in a very specific way because of objective situational and normative circumstances.

Thus, we must look at the situation of the emergency physician if we are to make judgments (or recommendations) about how this concrete person should act in a concrete situation of resource scarcity. The tension is between the *conceptual-general*, in which we draw our ethical conclusions, and the *concrete-individual*, in which we must act because we can only ever act at a particular space-time location and as the person we have come to be. No less a figure than Aristotle already knows about the importance of the concrete individual situation for ethical reasoning and therefore demands from the agent the virtue of *phronesis* (power of judgment; cf. Nichomachean Ethics VI, 1140a ff.), which can relate a generally made judgment to a concrete situation. But it is Cicero (or his Middle Stoic model) who explicitly focuses on the constitution of the agent himself. He does this in a writing that is

[13] I would argue that moral responsibility is not only the first factor in triage-decisions but also in other resource scarcities like the distribution of donated organs etc.

conceived as an instruction for his son, and can be considered the "original book" of applied ethics: *De officiis – On duties*.[14]

11.3.2 Four Roles

Cicero uses the word "persona", known from the Greek theater, which actually denotes a mask (off. 1107), and he states that we, as agents, wear four such masks: The first *persona* is the one we share with all our fellow human beings by participating in the one common reason.[15] In this "role", as the Ciceronian *persona* is generally translated today, we are able to engage in the general ethical reasoning ("ratio inveniendi officii") as we did above in relation to the scarce distribution of resources. But to see ourselves only in this role is just not enough. In a second *persona* we are physically and mentally determined, e.g. as fast or agile "persons", as rather humorous or serious. In the implementation of our actions we have to consider these "personal" aspects of our constitution. Who 'can't see blood' will not be able to become a doctor easily. The third role is imposed by chance or by the circumstances of the time ("casus aliqui aut tempus imponit"): Whether we grow up in a democracy or in a dictatorship is not in our hands,[16] and also the institutional structure to which one is subjected as a physician – at first anyway. This institutional structure can also include rules of prioritization, and to do justice to one's third *persona* then also means to comply with these rules and thus to take them into account in the same way as one takes into account general reason in the first person and the physical and mental constitution in the second person.

The fourth role deserves special attention, because it is not simply one to which the agent is subject, but in it a subject of action sets itself, as it were. It proceeds, as Cicero says, "from our will" ("a nostra voluntate proficiscitur"; off. 1115[17]). In the history of Greco-Roman philosophy, this is probably the first time that the subject also comes into view as normatively self-establishing and thus self-binding. This now opens up the possibility for a personal and professional ethos, which can no longer simply be deduced from general premises of the rational nature of man, but which realizes itself in its own task. The medical ethos represents such a setting made in self-commitment. Whoever as a physician sets himself the task of helping or assisting or healing refers on the one hand to given role models and thus to

[14] Cicero: De officiis, with an English translation by Walter Miller, Cambridge, Mass.: Harvard University Press (Loeb Classical Library), 1913, repr. 2005; Cicero: On Obligations. Translated with an Introduction and Notes by P. G. Walsh, Oxford: Oxford University Press 2008.

[15] "We must also grasp that nature has endowed us with what we may call a dual role in life. The first is that which all of us share by virtue of our participation in that reason and superiority by which we rise above the brute beasts." (transl. Walsh, p. 37).

[16] Thomas Nagel calls this "luck in one's circumstances." Thomas Nagel: Moral Luck. In: Moral Questions, Cambridge: Cambridge University Press 1979, 22nd printing 2018, 24–38, here 33 f.

[17] Walsh translates: "by our own choice" (p. 40).

aspects that originate from the third *persona*. But there is also something else, namely the acting out of free will within a form that one has given oneself. To give oneself such a form is ethically legitimate, because in doing so one does not give up rational binding and leaves room for arbitrariness, but on the contrary: By the choice of the form one binds oneself all the more, and the transgression of the normative form specification also represents a violation of the personal integrity.[18] Perhaps the example of a vegetarian can clarify the concern: There is probably no argument that could deduce compellingly from our general human nature (person 1) that one *must* be vegetarian. Nevertheless, there are good reasons (e.g. the horror of factory farming, the avoidance of animal suffering, climate concerns) that make it ethically obvious to choose a meatless diet. We then make such a decision in person 4. The one who has decided for vegetarianism should then also renounce meat consumption in the concrete situations, because otherwise he undermines his self-setting as a vegetarian – up to the point where it becomes ethically completely meaningless. Now here the consumption of animal meat is not to be made an issue, and presumably there are different views in terms of content, but it is about the analogy to the role-specific ethos of a doctor: Whoever decides for the profession of a physician or nurse decides for an ethos of helping – and decidedly not of judicial office or of a "moral judge". For this reason, the emergency physician in our initial example of the accident is justified in refusing to ask questions about the causation of the accident and the social consequences of his intervention. In his actions, she or he can and may start from the concrete emergency situation and act in his fourth role in a way that would not appear to be the correct course of action if viewed 'solely' from role One.

Cicero gives us the first formulation of role ethics. His considerations are certainly a bit rough and have not yet been adapted to individual professional roles. But that is not important in our context. It is crucial that modern role ethics can also be integrated into the model of order that Cicero has handed down to us.

We have two ends in the spectrum, which is fanned out by the model of the four personae: On the one hand, the unavoidable role of every human being as a human being, on the other hand, a role that is self-chosen but nevertheless becomes universally binding with the act of choice in the claim. Arthur I. Applbaum formulates the second point as follows: "Prescriptions can be *relative to the person* of the agent, in that particular persons, by virtue of their personal projects, commitments, and relationships, face particular moral reasons for action."[19] At the same time, Applbaum rejects Montaigne's insinuation that one can, to a certain extent, separate one's role obligations from one's actual person. "Montaigne thinks that the moral upshot of judgement's independence is that the person is not accountable for the vice and stupidity of one's role. Quite the opposite: not only is the deceitful lawyer or financier also a dishonest man, but he has the critical distance from whence to judge this.

[18] In that way it narrows down the gap between reasons and motivations.

[19] Arthur Isaak Applbaum: Ethics for Adversaries. The Morality of Roles in Public and Professional Life, (Princeton University Press) Princeton, NJ 1999, 64.

The detachable person is more, not less accountable for actions taken within a role, for the possibility of independent judgement blocks not only justification, but excuse." In Cicero's model: One never escapes one's first role as a human being.[20]

11.3.3 Institutional Protections and Its Limitations

The model of the four *personae*, which breaks up the one-dimensionality of ethical consideration from a general point of view of universal human reason, from which every right action is supposed to be derivable, must now, however, not itself be understood as a one-dimensional structure. Rather, within the structure, references are formed again. The medical ethos that someone chooses in the fourth *persona*, for example, also deserves institutional protection, which is guaranteed in the third *persona* – that is, the social and institutional role. It is therefore appropriate for professional associations to adopt codes of ethics that then lend normative force to the professional ethos from the institutional side as well. Such a position is taken, for example, by the "Ethical Principles of Health Care in Times of Armed Conflict and Other Emergencies"[21] adopted by numerous organizations with a high level of membership (such as the ICMM or the ICRC). These principles, which express a professional ethos, are rightly also relevant in a legal consideration where the law refers to the professional ethos. If, for example, the code states that "The primary task of health-care personnel is to preserve human physical and mental health and to alleviate suffering." (Art. 3), then an extra-moral purpose (the health of the patients entrusted to them) is specified, which may be used as a guideline from a moral perspective. For our case of the distribution of scarce resources, this means that prioritization should be based on medical considerations. However, it must also be kept in mind that such a set of rules should never be applied in the form of an algorithm. A human actor does not lose his (or her) practical power of judgment even in such a networked structure of role ties and must ultimately bring this power of judgment to bear. In many cases, this means that she or he can follow the given rules out of judgment. In some cases, however, this may mean that the path of strict rule-following is wrong.

After all, the professional ethos of the physician is not so clear-cut either. The best-known and most influential book on medical ethics by Tom Beauchamp and James Childress, "Principles of Biomedical Ethics", provides four principles which, in a process of specifying, must be determined more closely to concrete situations and, on the other hand, also want to be balanced against each other. This specifying and balancing is itself a non-algorithmizable task, for which the judgment of the individual person is required. The four principles are respect for autonomy,

---

[20] Applbaum, fn.19,. 75.

[21] World Medical Association, Ethical Principles of Health Care in Times of Armes Conflict and Other Emergencies, online: https://www.wma.net/wp-content/uploads/2017/02/4245_002_Ethical_principles_web.pdf [15/03/2023].

nonmaleficence, beneficence and justice. Beauchamp and Childress treat the distribution of scarce resources under the principle of justice,[22] but it remains quite underdetermined which forms of justice can or should be used in a distribution. It is assumed that it is a question of distributive justice, but as we have just seen at the beginning, distributive justice itself requires moral criteria. The classical criterion is desert. Everyone should get as much as she or he deserves. We come back to our initial problem: What is the measure of merit? Can fault reduce or limit merit? Do future expectations of a person's good actions increase his or her merit in the present? If one accepted such a prospective view of merit, this would already provide a connecting point to the – largely teleological – principle of beneficence. The point here in these considerations is not to solve a particular material problem of distribution, but to show the interlocking structure of ethical deliberation. Professional ethical standards influence the overall consideration quite decisively, but they are not the sole point of view from which a moral demand is made on the agent (here the physician).[23]

11.3.4 Excursus: Ex-Ante- and Ex-Post-Triage

Cicero's distinction of different personae is also helpful in grasping somewhat the intuitive difference between *ex-ante-* and *ex-post-*triage: For, of course, from the perspective of *persona 1*, that is, as it were, a "neutral" standpoint from outside the situation, it is not easy to see why a patient who is on a scarce ventilator and whose treatment is not very promising should take this ventilator even longer, when another patient with good prospects is coming in, but who can only survive if she herself is given a ventilator. Since there is only one respiration apparatus available, it seems obvious to perform an ex-post triage and to redistribute the respiratory equipment. But this is only part of the overall moral picture: The physician is also active in

---

[22] Tom Beauchamp/James Childress: Principles of Biomedical Ethics, 8. Edition, New York: Oxford University Press 2019, cf. "Allocating, Setting Priorities, and Rationing", pages 300–313. "Triage decisions usually appeal to medical utility rather than social utility. For example, disaster victims are generally sorted according to medical need. Those who have major injuries and will die without immediate help, but who can be salvaged, are ranked first; those whose treatment can be delayed without immediate danger are ranked second; those with minor injuries are ranked third; and those for whom no treatment will be efficacious are ranked fourth. This priority scheme is fair and does not involve judgements about individuals' comparative social worth." (p. 312). But note: The issue here is disaster victims, not victims of accidents with a human cause.

[23] The question of a moral criterion for the distribution of medical resources has recently also been discussed by Michael Gross: Military Medical Ethics in Contemporary Armed Conflict, Oxford: Oxford University Press 2021, 265ff. – Note that the moral responsibility account, as shown here, does not require punishment: It is a theory of the distribution of scarce resources. If there are enough resources for everyone involved in the emergency, everyone will be treated. – Therefore, the position put forward here does not presuppose "metaphysical freedom" but would also be defensible with compatibilism, because it is not a question of punishment based on free fault, but it remains a question of distribution.

another role (*persona 3/4*), and in this role she or he does not abandon his patient or leave her in the lurch (cf. Dietrich 2021, 94). There remains a tension: There is no algorithm by which it could be clearly calculated when role 1 or role 2 "wins", i.e. under which circumstances a patient can ultimately be taken off the ventilator in order to save another person's life. The action-theoretical distinctions between doing and allowing are weighty, but often ambiguous or can also be undermined by certain mechanisms (e.g., by the fact that the machine regularly switches itself off and must be actively switched on). It is the judgment of the individual human being who has to make a final decision here between the role claims that tips the scales.

11.3.5 The Complexity of Roles with Military Medical Personnel

So far, I have kept the conceptual tools for our considerations quite reduced. It would certainly be possible to take a closer look at the subject on the basis of the concepts of conscience, virtue or responsibility – social roles usually generate specific social responsibility. In addition, there is the difficulty that in ethics we hardly know how to deal with the factor of time – especially where time becomes history. This is essential in view of the professional-patient-relationship. A physician certainly has a greater responsibility (to use this term after all) towards a patient with whom he or she has an already longer treatment history. But is this a factor that should not be weighed against another factor, such as medical necessity? What would we say if, in the COVID-19 vaccination campaign of 2021, primary care physicians who received vaccine to distribute to their patients informed only their "regular" patients about it, and not occasional patients, but who, from the standpoint of need, needed the vaccine even more urgently?

Military medical personnel now face yet another very unique situation that adds to the moral challenges. Belonging to a certain soldierly community, eating together with people, having conversations, sharing life problems leads to a bond that cannot be neutralized by decree. The claim of completely neutral medical action between two or more parties to a conflict, which International Humanitarian Law in a sense places on military medics, is just that: a claim. But that does not mean that there may not be other claims that also have some moral justification. Nor does it make sense to completely fail to recognize the emotional and motivational resources for ethical action: Proximity is one such resource. It would be a form of overmoralization to criticize a military medic for giving medical preference to his comrades over wounded soldiers of the opposing conflict party in certain very tough conflict situations. But it is a different matter if one were to establish a general institutional obligation to give preference to one's own wounded over those of the enemy. In this case, the military medical service would very quickly become embedded in a functional arrangement that instrumentalizes it for the war aims of one's own party. But this would undermine the medical ethos, which, as we said above, would turn the task of normative institutions, which are supposed to protect professional ethos, into its opposite. In the language of Cicero's four *personae*: Even if, from a 'person One point of view', one's own war aims are the justified ones and those of the enemy are unjustified, this does not cancel out the moral significance of roles 3 (social role which is institutionally deposited) and 4 (the self-chosen role as a physician).

The professions of physician or nurse, especially in emergency medicine or in the context of armed conflict, means that one may have to act under enormous ethical challenges. The four *personae* model, if taken seriously, does not provide a handhold for moral exoneration. It is not a matter of people, under pressure to act, choosing whichever of the four roles suits them for particular reasons. A military medic who treats only the members of his own conflict party because in his first role as bearer of general reason he comes to the conclusion that only his own party to the conflict is waging a just war and therefore his comrades must receive full protection and treatment, while the enemy would have no right to fight at all and is thus liable for his own wounds, fails to fulfill his other roles, which are also morally weighty. The same thing holds true for biased distribution of scarce resources. Rather, the model of the four *personae* illustrates the complex trade-offs that must be made in practical situations and, in this way, even raises the stakes. But at the same time it represents a first and classical attempt from the philosophy of the Stoics to do justice to this complexity in our practical considerations by means of a conceptual model. Real relief is offered only by an early recognition of the scarcity of resources, so that it can be largely avoided. Of course, a society cannot put all its resources into medical resources and must also weigh things up politically wisely here, but especially when it comes to military deployment scenarios, planners should be particularly careful not to put their medical personnel into morally oppressive and thus also psychologically hardly bearable situations.

11.4 Conclusion

Karl Marx, in his "Critique of the Gotha Program" (1875), described the idea of "just distribution"[24] as insoluble, because every idea of justice is basically based on an ideology. In fact, according to Marx, legal concepts spring from economic relations. The considerations made here seem to prove Marx right to a certain extent: If there is a scarcity of resources essential for survival, only on the basis of ideological presuppositions can a clarification of the question of distribution seem possible. However, one "ideology" in particular has been singled out here, namely that which assumes human agency and moral responsibility. In a certain sense, this is the "ideology of ethics" in general. People are responsible for their actions; these actions

[24] Karl Marx: Kritik des Gothaer Programms. In: Karl Marx/Friedrich Engels: Werke, Band 19, Berlin: Dietz Verlag 1982, 11–32, 18.

have consequences, and these consequences are also to be attributed to the people who are capable of acting. In this respect, a person acting in a situation of scarcity of resources, which she or he himself or herself has brought about, must also be the first to bear the consequences of this scarcity. But this moral person (Immanuel Kant speaks of "homo noumenon") is not the empirical human being that we – and precisely physicians in emergency situations – have concretely before us ("homo phaenomenon").[25] Formal ethical requirements (e.g., in the allocation of resources) for medical personnel have to be fulfilled. Therefore, it has been suggested here to recall the first text in the Western tradition of ethics in which the distinction of role-specific duties was made, namely Cicero's "On Duties". Of course, this does not conclusively answer the material question of what criteria should now guide action within a particular role. However, there are certain basic principles, e.g., the role-specific bond between physician and patient, which precludes the permanent reassignment of patients in the ex-post-triage process, or public normative definitions in the form of professional codes of ethics. In this model, however, the agent himself stands in a plurality of roles and must therefore also have the power of judgment to know in which situations she or he may need to transcend a particular role. The power of judgment, however, can only be trained (above all also on real and hypothetical, also literary examples) and cannot be algorithmized by further norm specifications.

References

Applbaum, Arthur Isak. 1999. Ethics for adversaries. In *The morality of roles in public and professional life*. Princeton, NJ: Princeton University Press.

Beauchamp, Tom, and James Childress. 2019. *Principles of Biomedical Ethics*. New York: Oxford University Press.

Cicero. 1913, repr. 2005. *De officiis*, with an English translation by Walter Miller. Cambridge, MA: Harvard University Press (Loeb Classical Library).

———. 2008. *On obligations*. Translated with an Introduction and Notes by P. G. Walsh. Oxford: Oxford University Press.

Dietrich, Frank. 2021. Medizin am Limit: Wie umgehen mit Versorgungsengpässen in der Pandemie? In *Nachdenken über Corona. Philosophische Essays über die Pandemie und ihre Folgen*, eds. Geert Keil/Romy Jaster, 84–97. Stuttgart: Reclam.

Dufner, Annette. 2021. *Welche Leben soll man retten? Eine Ethik für medizinische Hilfskonflikte*. Berlin: Suhrkamp.

Frowe, Helen. 2014. *Defensive Killing*. Oxford: Oxford University Press.

Gross, Michael. 2021. *Military medical ethics in contemporary armed conflict*. Oxford: Oxford University Press.

[25] In Kant's view humans are to be understood as "citizens of two worlds": As beings of the senses ("homo phaenomenon"), humans are subject to the laws of nature and can be examined by empirical means, as beings of understanding ("homo noumenon") they belong to an intelligible world. As intellectual beings, human beings ask for the principles of their actions and cannot be examined by empirical means. The central source for knowledge of the quality of the *homo noumenon* is basically the introspection of the person concerned ('heart-' or 'soul-searching').

Mannino, Adrianno. 2021. *Wen rette ich – und wenn ja, wie viele? Über Triage und Verteilungsgerechtigkeit.* Stuttgart: Reclam.
Marx, Karl. 1982. *Kritik des Gothaer Programms.* In Karl Marx/Friedrich Engels: *Werke*, Band 19, Berlin: Dietz Verlag, 11–32.
McMahan, Jeff. 2009. *Killing in war.* Oxford: Oxford University Press.
Nagel, Thomas. 1979. 22nd printing 2018. Moral Luck. In *Moral Questions*, 24–38. Cambridge: Cambridge University Press.
Quante, Michael. 22014. *Menschenwürde und personale Autonomie. Demokratische Werte im Kontext der Lebenswissenschaften.* Hamburg: Meiner.
Walzer, Michael. 2015. *Just and unjust wars.* New York: Basic Books.
Williams, Bernard. 1981. Moral Luck. In *Philosophical Papers 1973–1980*, 20–39. Cambridge: Cambridge University Press.
World Medical Association et al. n.d. *Ethical Principles of Health Care in Times of Armed Conflict and Other Emergencies.* https://www.wma.net/wp-content/uploads/2017/02/4245_002_Ethical_principles_web.pdf

Bernhard Koch is the deputy director of the Institute for Theology and Peace in Hamburg, an associate professor at the Faculty of Theology of the Albert Ludwig University in Freiburg, and a lecturer in philosophy at the Goethe University in Frankfurt am Main.

Chapter 12
Facing Death: An Ethical Exploration of Thanatophobia in Combat Casualty Care

Erika Ann Jeschke, Hannah R. Martinez, Eleanor M. Choi, John Dorsch, and Sarah L. Huffman

Abstract This paper is going to explore the adverse effects of exposure to combat death on medics' holistic well-being, which, if ignored could decrease individual readiness and negatively impact the mission. We rely on the experience of United States Air Force Special Operation Surgical Teams (AF SOST) whose exposure to mass casualty scenarios in austere environments could serve as approximations of conditions of future battlefields. Over the past two decades, the ability to deliver advanced medical care on and off the battlefield along with a rapid casualty evacuation platform has allowed for unprecedented survival rates exceeding 90%. While laudable, these medical achievements have also set up a casualty management paradigm in which medical decision-making singularly focuses on life-saving care. Confronting peer-peer adversaries in large scale combat operations (LSCO) on a multi-domain battlefield will make casualty management frighteningly more complex by introducing major infrastructural, personnel, and resource constraints. When considering the high number of casualties expected in LSCO alongside enormous limitations in medical resource and resupply capability, the current casualty management paradigm will not be sustainable. The resource constrained environ-

E. A. Jeschke (✉)
Senior Lead Scientist, Parsons Corporation, Navy Medical Research Unit-Dayton, Dayton, OH, USA

H. R. Martinez
Anchorage, AK, USA
e-mail: hannah.r.martinez4.mil@health.mil

E. M. Choi
Georgetown University, Washington, DC, USA

J. Dorsch
UNC Southeastern Medical Center, Lumberton, NC, USA

S. L. Huffman
Director of BioBehavioral Health Research, David Grant Medical Center, Fairfield, CA, USA
e-mail: sarah.l.huffman.mil@health.mil

S. M. Eagan, D. Messelken (eds.), *Resource Scarcity in Austere Environments*, Military and Humanitarian Health Ethics,
https://doi.org/10.1007/978-3-031-29059-6_12

ment in LSCO will shift medical decision-making away from a singular focus on life-saving care to triage, which hinges on the ability to determine futile medical interventions—a skill that has been lost in the past two generation of combat medics. As such, a broad ethical challenge that arises in preparation for LSCO is the need to set new expectations concerning dying and death. However, medical decision-making focused on death and dying has not been explicitly addressed in military medical training, research, or policy. Relying on a body of literature known as terror management theory (TMT) and ethnographic data from our study with AF SOST medics, we are going to argue that it is important to expose medics to death in mass casualty and triage training.

Keywords Thanatophobia · Combat casualty care · Battlefield medicine · Ethics

12.1 Background

This paper will explore the adverse effects of combat death exposure on medics'[1] holistic well-being, which if ignored could decrease individual readiness and negatively impact the mission. We rely on the experience of United States Air Force Special Operation Surgical Teams (AF SOST) whose exposure to mass casualty scenarios in austere environments could serve as approximations of conditions of future battlefields. Over the past two decades, the ability to deliver advanced medical care on and off the battlefield along with a rapid casualty evacuation platform has allowed for unprecedented survival rates exceeding 90% amongst injured American service members. In fact, both academic and casual discourse in military medicine speaks of survivability rates as high as 98% if a casualty arrives alive to a combat hospital (Butler et al. 2015). This medical achievement is not only impressive and laudable, but also right and good to acknowledge as one of the exemplary successes of modern military medicine. However, there are also unexpected side effects to achieving great success. The challenges and consequences of the downsides of success can easily be eclipsed, if academic and cultural discourse primarily focus on the glory of victory. One of the downsides of a casualty management paradigm focused on survivability is that death and dying are treated as hypothetical and abstract possibilities to be avoided at all costs. With advances in technology, death itself becomes unacceptable, thereby breeding a learned helplessness into medics on the battlefield when they are faced with its inevitability. If death and dying are not directly addressed in training, they could cause medics to experience unconditioned negative stimuli for which they are not prepared to manage.

[1] In this paper, we will use the word medic as a general term for all military service members involved in combat casualty care. While we will be focusing on those who provide medical interventions and the military medical culture, it is important to note that the military chaplain is also a member of the combat casualty care team.

Clearly, battlefield medicine is dynamic and contingent on the context of fighting. The current casualty management paradigm and the survivability rates that it supports are predicated on a resource rich environment composed of: adequate numbers of highly trained and interdisciplinary medical professionals, technologically advanced equipment, specialized supplies, and sufficient time enabled by a rapid areo-evacuation system. Looking to the future fight, military strategists are discussing how to prepare for large scale combat operations (LSCO) with peer-peer adversaries on a multi-domain battlefield. This battlefield will be defined by anti-access and area denial, which will introduce major infrastructural, personnel, and resource constraints. In specific, there will be no capability to support large medical teams, provide medical resupply or quickly evacuate casualties. When considering the military capabilities of a peer – peer adversary, medical experts anticipate upwards of 10,000 wounded per month (Davis and Rasmussen 2019). With only limited medics and supplies, casualty management and medical decision making will be frighteningly more complex. Considering the high number of casualties expected in LSCO, alongside enormous limitations in medical resource and resupply capability, it seems obvious that the current casualty management paradigm will not be sustainable.

Unfortunately, singularly focusing on life-saving medical interventions has set up a military, medical, and ethical fiction that not only depicts war without death, but it forwards a normative goal of achieving this political reality. Since few medics have faced an austere battlefield with limited resource and resupply capability, the challenges of this fiction have gone largely unnoticed and undiscussed by the broader military medical community. One of the practical implications of a casualty averse military culture (Lacquement 2004) and a death averse medical culture (Kübler-Ross 1969) is that training for complex medical decision making amidst mass casualty has become emaciated because it ignores death and dying as part of the triage calculus (Jeschke 2019). When considering a resource constrained environment like LSCO, mass casualty scenarios will be defined less by survivability rates and more by death and dying. As such, triage will shift away from a singular focus on sorting medical resources to provide life-saving interventions to primarily focus on the management of salvageable casualties. Such an environment will also likely be defined by reverse triage scenarios where the least injured casualties will be treated first to return them to the fight. Under such conditions, medics will be forced to make ethical decisions about withholding medical intervention and how to best ethically attend to death. Put bluntly death and dying will return from the shadows of abstraction and be present as both a military and medical reality on future battlefields.

If the military does not address its culture of death denial, combat casualty care providers will be ill-equipped to engage medical decision making amidst the types of mass casualty scenarios anticipated in LSCO. Shifting away from death denial is not a small challenge, as it demands major change that reaches down to the deep roots of both the Western military and medical cultures. However, we believe that it is necessary to prepare and support medics to engage in ethical and effective triage in LSCO. If death and dying are not directly addressed in training, these human phenomena will likely cause medics to experience unconditioned negative stimuli for which they are not prepared to manage. Such stimuli could have an immediate negative impact on mission success and long-term negative consequences to well-being. Furthermore, the backlash of not upholding the implicit cultural-political normative goals entailed in life-saving medicine could become an inordinate burden on the medics. The future reality of LSCO presents an ethical challenge—the need to set new cultural-political expectations concerning dying and death in combat casualty care.

At the broadest level, this paper will address the phenomenon of death aversion in combat casualty care using an interdisciplinary approach. More specifically, our goal is to uncover some of the ways exposure to death and dying impacts medics who have tended to mass casualty scenarios in austere, resource-depleted environments. We will rely on two primary sources of data. To frame our ethical reflection on medical decision making and death aversion, we turn to body of literature in social psychology known as Terror Management Theory (TMT). Our applied work will rely on data that has been gathered from our empirical study entitled, *The Impact of Catastrophic Injury Exposure on Resilience in Special Operations Surgical Team Members*. Since this is a small community of military medics, all ethnographic data has been deidentified in a two-fold process. First, we removed all locations, names, genders, ranks, and medical specialties. Second, we created composite quotes that capture thematic nuance, but avoid using words from one single participant. Data includes four participants from each of the six medical specialties represented in a AF SOST: general surgeon, emergency physician, nurse anesthetist, critical care nurse, respiratory technician, and surgical technician.

To be clear from the outset, our analysis is not meant to suggest that the current casualty management paradigm is unethical or that military medical training ought not focus on survivability. We hope to paint a more nuanced picture that illustrates areas of expansion and improvement that might help manage necessary change in preparation for future battlefields. We also believe that this direction of research and training will positively impact medics' readiness for the future fight as well as their long-term holistic well-being. When thinking about how medics ought to manage combat casualties in austere environments as well as what obligations of care are due to medics upon returning home, we are going to argue that pre-deployment medical training needs to include mass casualty scenarios that rigorously familiarize medics with the complex decision making involved in allowing some casualties to die. We believe that such training is essential to support ethical and effective medical performance on future battlefields. We also believe that enhancing training in this area would support medics' ability to negotiate complicated ethical decision making when it comes to triage. As an interdisciplinary team, we aim not only to open a new space of ethical reflection about military medical decision making in resource depleted environments, but also influence applied military health research, such that the military community begins to acknowledge the burdens associated with combat casualty care. We also believe that such acknowledgement is an essential staring point from which to develop meaningful resources to support medics' post-deployment healing and reintegration.

Our argument will unfold in five sections. First, we present a conceptual model of how death aversion functions in both the military and medical cultures. Second, we illustrate how the current paradigm of casualty management establishes implicit normative expectations about what counts as a good medic and good combat casualty care. These normative expectations, in turn, shape the medic's identity and influence medical decision making. Third, we briefly review some of the overarching principles of TMT to highlight how death aversion in large groups impacts medical decision making. Fourth, we highlight some of the emerging themes in our ethnographic project with AF SOST medics to illustrate the complexity of medical decision making in resource depleted environments. This section will establish the impact of death and dying on medics. Finally, we lean into potential ways of shifting cultural expectations around dying and death in combat casualty care training specifically focused on death as an outcome.

12.1.1 Conceptual Diagram of Death Aversion in Modern Military and Medical Cultures

To begin we present a visual diagram that illustrates how the modern western military and medical cultures implicitly relate to death as a human phenomenon. Physiological death is a negative human phenomenon to be completely avoided whereas physiological life is a positive human phenomenon to be completely embraced. In other words, there is no possibility for a negative relationship to life or a positive relationship to death in the military and medical cultures (Diagram 12.1).

The reader will note, both the military and medical cultures posit that the relationship between life and death is linear when represented on a graphic plane. In this model, *death* is pathological and pertains only to the biological death of a casualty. Inextricably associated with a deleterious relationship to life, a total absence of any relationship to death is considered healthy. Therefore, death resides in the negative quadrant of the graphic plane. Similarly, the only possible option for engaging in a healthy relationship with life resides in the positive quadrant and is defined by total absence of physiological pathology. As you can see, this model leaves no room for a salubrious relationship with death or a deleterious relationship with life as a normal part of human existence. While movement is possible between life and death, movement must always go away from physiological death towards physiological life. In relationship to death, any other movement is always normatively problematic in the military and medical cultures.

When considering this model, it is easy to see how the singular focus on lifesaving medicine is entangled in implicit normative expectations about life and death. Death of an American Service Member is to be avoided at all costs, whereas life is to be always promoted. Health and living solely pertain to the physiological functioning of the human being without consideration of the context in which that life takes place. In other words, it does not consider that a negative relationship to

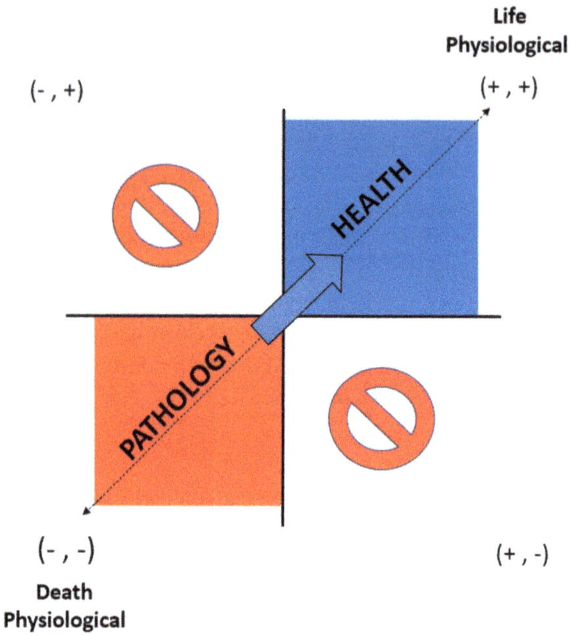

Diagram 12.1 Military and medical cultural relationship to life and death. (Constructed by Erika "Ann" Jeschke, Hannah R. Martinez, & Elle Choi)

life or a positive relationship to death could be entail salubrious aspects of holistic well-being. Said differently, within the military and medical cultures there is no way to interpret negativity in life or positivity in death as a rich source of meaning that supports the flourishing of a medic or the casualty in the medic's care. One specific example would be that it forecloses any possibility to narrate a good combat death for the warfighter or the medic. This will be further addressed in a subsequent section of this paper.

The normativity of death aversion in the medical culture is well articulated by Nadine Ehlers and Shiloh Krupar in their book *Deadly Biocultures.* They suggest that, "Life-making remains the dominant political paradigm in U.S. biocultures today … Life is maximized, optimized, and prolonged via a range of strategies and technologies that affirm life. Within this paradigm, life is framed as that which must be constantly tended and relentlessly avowed: death must be disavowed. Such harnessing of life is promoted as a universal good." (Ehlers and Krupar 2019, p. 161). It is easy to think of countless ways in which the modern medical culture is woven together with life-making processes that focus on sustaining the physiological functioning of a human being.

In contrast, the reader might find it odd to suggest that the Western modern military is also caught up in death aversion considering that combat operations involve deadly force. However, the cultural entanglements of death aversion in the United

States military can be found in a rigorously researched in military strategic studies when looking at the phenomenon of casualty aversion. Richard A. Lacquement Jr. well articulates the challenge of casualty aversion in military and political decision making saying, "Avoidance of casualties is an unassailably desirable objective. It is precisely the natural nobility of the argument that makes it susceptible to misuse in the policy-making process, potentially leading to ineffective or inefficient choices." (Lacquement 2004, p. 40). Casualty aversion is not just illustrated in military and political decision making, but it also impacts how service members perceive their own sense of sacrifice, what counts for good sacrifice, and what ought not to be sacrificed. To specify, sacrifices are to be made up to a point, but that point ends with death of a United States Service Member. As Sebastian Kaempf suggests in his book *Saving Soldiers or Civilians*, ancient military culture included a willingness embrace death in its understanding of battlefield sacrifice. This willingness to confront death in combat infused ancient military culture with rich meaning. However, since the Vietnam conflict the United States military has been unwilling to include death as part of its cultural understanding of "right" and "good" sacrifice. Death embrace was replaced by death aversion as the culturally normative relationship to United States casualties (Kaempf 2018). As such, death aversion and casualty aversion are mutually reinforcing cultural entanglements that strongly influence military understanding of what counts as good medicine and the good medic—both are singularly focused on life-saving medical interventions that uphold the normative goal of achieving war without death for United States Service Members. To establish a more vivid context in which this ethical normativity exists, we turn next to empirical data.

12.1.2 Current Cultural Paradigm of Casualty Management

The dominant ethos of combat casualty care is do everything possible for as many people as possible because military medicine is about the business of saving lives. As one AF SOST medic stated, "Our job is to battle and defeat death." As mentioned in the introduction, casual discourse within military medicine references a 98% survivability rate after battlefield injury. This statistic is rarely qualified and highly open for interpretation, especially in terms of who and what counts as casualty. Culturally, 98% has become a symbolic reference point for good medicine in combat casualty care. In other words, the implicit normative thrust of military medicine is the ability to provide life-saving medical interventions that not only uphold, but also increase this symbol of goodness. Normatively linking good medicine to physiological survivability, also circumscribes what counts as care. In other words, care becomes culturally synonymous with medical interventions and does not necessarily entail notions of kindness, comfort, or concern for casualties in any way other than physiological survival. Additionally, when care is circumscribed as medical interventions that are meant to maintain survivability after injury, the burden medics carry in relationship to providing medical interventions for casualties is not included within the scope of care.

Good medicine as survivability is evidenced in policy initiatives such as the zero preventable deaths after combat injury (National Academies of Sciences & Medicine 2016). The goal of this initiative is to improve military trauma care, which, on face value, seems commendable at best and innocuous, at worst, when thinking about military medical research and training. However, research and training are decontextualized from the myriad of challenges present on an austere battlefield and can never recreate all the variables that will be involved in medical decision making. Furthermore, such initiatives are based on the assumption that battlefield medicine is a static reality and will continue to be highly resourced.

The context of austere medicine in relationship to survivability and casualties is highly dynamic and, by definition, resource depleted. The following narrative illustrates some of the complexities of casualty management in a resource scarce environment:

> From the first day the flood gates just opened and it was a steady flow of casualties 24 hours a day seven days a week. Trucks would show up with a bunch of bodies in the back. On our busiest days we would see about 20 trauma patients. In that period, we had 19 mass casualty events. About 750 total casualties in a seven-week period with 6 of us working almost 24 hours a day. All of us wanted the best outcome for the patients. We did the best we could with the resources we brought, but they ran out. We didn't have exam gloves. We were using lunch lady gloves the locals brought us, but your finger goes through lunch lady gloves. We were putting in chest tubes with our fingers inside the person's chest. I know that would be scrutinized by other people. They would question ... But what else do you, let the person die? We ran out of everything. I remember one time when we had 20 casualties come at once. One casualty was shot in the belly and bleeding from the pelvis. We went to the Operating Room because the others seemed stable. Once we got started with surgery, we were not successful in controlling the bleeding, then more casualties started to arrive. A team member came in and told us that they had at least two other casualties that both needed an immediate operation and there was not enough blood to save all three casualties.

Furthermore, these ideals and aspirations of improvement insinuate themselves into cultural ideals and normative expectations about what it means to be a good medic in relationship to preventable deaths. While there are factual standards of what counts as preventable death (Janak et al. 2018), practically speaking, the 2% is an interpretive heuristic against which medics judge their own success or failure. In particular, interpretation of what counts as preventable death for the "good medic" is seen as achieving 100% survivability in all contexts under all conditions, even amongst complex medical scenarios as described above. To highlight this point, deciding to withdrawal life-saving medical intervention in the previous narrative was interpreted by one of the team members as follows: "We consciously chose to let this guy bleed out and die, so that's really hard. It feels almost like we killed him, and in the strictest sense of the word that's exactly what we did." Obviously, readers familiar with the ethics of withholding and withdrawing of care will quickly realize that this sort of medical decision making does not fall within the category of killing.

However, the 2% standard of perfection gets culturally interpreted by medics in such a way that the expectation is that no one should die under the most inordinate circumstances. In specific, medics view their job as closing the gap between 98% and 100%. If someone dies on their watch, the mythos of the 2% preventable death hits their psyche hard.

It is also interesting to note how these culturally normative standards of survivability influence and impact training. Out of all of our participants, only one participant explicitly received preparatory training for the type of austere mass casualty triage that would be faced as an AF SOST medic. More interestingly, that training did not come from military training, but from pre-hospital care that was done as a part of civilian emergency medical training. Almost all participants suggested that training was either insufficient or too abstract for the reality that they were to face. The type of triage training was explained in the following way that confirms how death remains an abstract phenomenon that occurs on a hypothetical battlefield with controllable outcomes:

> Noteworthy experts gave seminars about triage and mass casualties. The goal was to teach the right way to make medical decisions in these circumstances. They showed all this conceptual modeling, but the teachers had never engaged medical decision making in triage or mass casualty event. Triage is a dynamic thing. The way it gets taught you is that it is a fixed rigid thing. The patient with X situation gets X, the patient with Z situation gets Z. When you are a very small team and you have several dozen patients how do you really truly triage. The first time we were faced with the brutal reality of triage, we realized we had really hard decisions to make and they had to make it quickly. It was frustrating to learn that we were only a few people and only capable of doing a certain number of things. The more we understood what triage really is—that it is a very dynamic fluid thing based on the environment we are in—the more we started to realize survivability is situation dependent. It was very challenging and agonizing thing to get into because we had to make decision death. So, someone that is survivable in some situations is not going to be survivable in others. It is very challenging learn the really difficult balance of triage and medical decision making amidst a mass casualty because we had to watch people die that under other circumstances we could have saved.

Training in the abstract using death as a hypothetical that can be calculated through rational analysis, leaves medics unprepared for the complexity of ethical decision making and visceral encounters with death and dying. It also builds in an assumption that rational training is sufficient to prepare them for complex medical decision making in real time scenarios. As such death and dying become static realities that are trained to be controlled not to be encountered. Unlike other forms of medical or tactical training that function more like Aristotelean phronesis—practical wisdom that is formed through habituation of skills—triage training is performed in a more as an analytic, abstract ethical process. However, the dynamic environment in which triage takes place amidst a mass casualty in austere combat environments seems more aligned with phronesis that seeks to make the right decision, at the right time, with the right people, in the right way. In other words, triage is bound up in the immediate contextual details in which decision-making occurs.

Unfortunately, the nuance of survivability is not learned until medics come face to face with the limitations of abstract triage training. As such these types of training are predicated on the cultural ideal that medics can achieve 100% success in preventing deaths if their medical and rational skills are sufficiently honed and applied. The interpretive vacuum of interpretation into which meaning is made about what counts as casualty, survivability, preventable death as well as good medicine and medic is poignantly illustrated by the following comment:

> I will tell people 'if you can get someone to us alive. We will save them. That someone is an allusive term … If that someone happens to be an American, then we have a system to save them … The thing that keeps me in my uniform is knowing that if there is an American soldier hurt anywhere in the world, the United States government will do everything and utilize every resource to get to that one soldier on the front-line and get the soldier home alive.

Medics in our ethnography rationally understood that 100% survivability was not a realistic outcome of combat casualty care and acknowledged, in the abstract, that some people would die. However, when it came to their own self-assessment of good medical care, every medic believed they had done something wrong if a casualty died while they were performing a life-saving medical intervention. As was explained about the AF SOST mission:

> We don't exist for the average casualty. We exist for the one-offs, if we're doing things right, if we're going into the environments we're going into, we might get one chance to get it right. You can't mess it up or it's 100%. We had one American that didn't survive. We definitely don't think about the hundreds of lives we saved. We think about the one we didn't save. The scene is played over and over and over again. What could we have done differently? How do we never ever make the same mistake again?

Our initial data suggests that medics function in a cultural pressure cooker established by the current paradigm of casualty management. Even in the most austere environments, medics attempted to embody medical perfectionism to avoid. The 2% statistic is the implicit standard against which medics emotionally and experientially measure the value of surgical intervention provided. When 100% success in preventing death was not achieved, medics expressed a sense of guilt and failure even though they readily acknowledged that not all conditions would allow them to prevent death as a medical outcome.

The symbolic value of 2% has an almost mythical status in the culture of combat casualty care and holds enormous sway over medics. As such, the implicit normative force of 98% survivability and 2% culturally counting as preventable death strongly shapes not only what counts for a good medic, but also the identity of medics. Said differently, medics start to take on an identity of heroic combat life-saver. While there are undoubtably many goods that come from this identity, it is also saturated in death aversion. Next, we turn to TMT to show how a death averse military medical culture not only shapes the medic's heroic life-saving identity, but also influences medical decision making when thinking about triage and mass casualties in a resource depleted environment.

12.1.3 Terror Management Theory and Medical Decision Making

TMT is based on the research of Jeff Greenberg, Sheldon Solomon, and Tom Pyszczynski. This theory is derived from Ernest Becker's book *The Denial of Death* in which he explores both Kierkegaard and Freud's theoretical understanding of *thanatos* in relationship to human behavior. Becker suggests that human beings primordially function from a sense of *thanatophobia*, or fear of death. This fear causes individuals to form exclusive groups in which they are able to construct heroic and idealistic self-concepts to feel impervious to death (Becker 1997). Greenberg, Solomon, and Pyszcynski picked up on Becker's idea in their work as social psychologists and spent decades studying human behavior to better understand how individuals make decisions when they are directly confronted with images or thoughts of human mortality. In specific, they evaluated how mortality salience—confrontation with mortality—led to group conformity, obedience to a charismatic-autocratic leadership, self-serving biases, aggression against and prejudice of individuals outside of the chosen in-group. Their major findings are laid out in the book "*The Worm at the Core: On the Role of Death in Life*" (Solomon et al. 2015).

After 30 decades of research and over 400 studies on mortality salience, TMT suggests that if death denial exists, then reminding a person of his/her individual mortality increases positive reactions to those who validate aspects of that individual's heroic identity and worldview. In contrast, physical and/or verbal aggression is shown towards individuals who act contrary to the heroic identity or worldview of the in-group. Greenberg, Solomon, and Pyszcynski also found that mortality salience increases striving for individual self-worth and acknowledgement from fellow like-minded group members. These behaviors reinforce the individual's sense of heroic self-assessment as well as increasing positive reactions to heroic figures that serve to uphold the cultural assumptions of the in-group. Taken together, mortality salience leads to displays of personal boldness that are meant to sustain an individual's status within the in-group by garnering positive affirmation from the group. Mortality salience also increases devotion and adherence to charismatic authoritarian leaders who provide overly simplistic answers to complex social dilemmas. In specific, these leaders reinforce narratives that describe an in-group in grandiose terms and describe the outgroup as socially problematic (Greenberg and Arndt 2011). At its worst, TMT suggests that death aversion could eventually lead to a culture that ignores criminal behavior in favor of protection of group identity and adherence to authoritarian leadership.

Our ethnography has not shown any problems with negligent medical or military behavior. Quite the contrary, the AF SOST medics have shown enormous generosity, concern, and kindness when caring for all casualties in the under the most compromised medical and ethical conditions. The following narrative illustrates the virtue with which AF SOST medics have ethically navigated various tension of combat casualty care:

> The first time I saw an American killed, after I worked on that service member, I turned to the person that killed the service member. Obviously, I'm human so there was some negative emotion. The anger and the emotion that goes into trying to take care of an enemy is enormous. There's always that part of me, as a human being, that is tempted to mistreat the enemy or take revenge, but there's another part of me that realizes "If I act cruelly toward this casualty, I validate every negative perspective this person was ever taught." In my career I've taken care of people that I know have killed Americans. I've taken care of people I know have killed women and children. It doesn't come without its moral dilemmas. When I take care of an enemy that I know has killed somebody on our side, I like to put myself in that person's scenario. This person is completely helpless, probably in the most vulnerable position they've ever been in in their life. They're hurt, they're in the hands of the enemy, they're naked on a table, and they're looking up at me. I guarantee you, because I've seen it, if I take good care of them, they wake up, they do well, they get treated with respect and compassion, then it changes the dynamics. The best weapon I we have is to give every casualty compassion and care. Even we may be angry and even though there's that part of us that is trained to be American warfighters, that the bigger part of our humanity is the bigger part of war and the bigger part of the battle being decent to the enemy. Nothing is to be gained from hurting somebody in that scenario. Nothing is to be gained from getting revenge in that scenario. Everything is to be gained from changing the way somebody thinks. Maybe you do, maybe you don't.

Of the 24 medics interviewed, all have described a similarly profound experience providing compassionate and dedicated care to all casualties as well as an enormous sense of loss over the death of any casualty.

Although we have seen enormous charity expressed by AF SOST medics when tending to casualty, TMT also suggests that death aversion could lead to decreased mission effectiveness in term of various power-plays between medical teams and other military teams. In addition, death aversion could potentiate challenges in medical decision making when thinking about triage and mass casualties in austere environments. Both of these situations have come up in our data collection. Many medics have described experiences where the AF SOST struggles to build rapport with the Special Operations Forces team (Operational Detachment Alpha/ODA) with whom they work and are trained to support. Interestingly, the conflicts between the ODA and AF SOST are described as an identity conflict that manifests as tension between both groups' heroic identity: the AF SOST medic's identity as combat life-saver and the ODA's identity as ultimate fighter able to swiftly subdue the enemy. Death aversion manifests in the data as group-fights around different, but strongly fortified heroic identity structures—the ODA's ability to subdue the enemy with incredible tactical precision and the AF SOST's ability to save life with medical precision.[2] This continues to connect to the Western medical culture's singular

[2] When presenting data to various audiences as well as during various field observation, this infighting has been noted across various forward surgical teams. For instance, members of the United States Army Forward Surgical Team (FST), the Joint Special Operations Command's (JSOC) Joint Medical Augmentation Unit (JMAU), and the United States Air Force Ground Surgical Teams (GST) have engaged in dismissive commentaries the value of the other surgical teams worth. This suggests that thanatophobia plays into identity conflicts across medical teams. It also suggests in-group exclusive behaviors in reference to combat casualty care and understanding who is best prepared, equipped or capable of managing combat casualties. While this is not the topic of our

focus on life-saving interventions and the American military culture's singular focus on reducing Service Member casualties. As such, death aversion is the common cultural foundation which reinforces both the AF SOST and the ODA's in-group heroic identity. This tension was somewhat surprising to our research team because the ODA is trained to inflict mortal harm and AF SOST medics are trained to be medically agile and tactically prudent. We had originally anticipated that tactical training would make AF SOST medics more amenable to ODA. In contrast, there seems to an enormous taxation on AF SOST medics because their battlefield medical prowess is de-valued or seen as a liability to the military mission. As such, AF SOSTs spend a great deal of effort in trying to manage the ODA's impression of surgical team's ability to be an asset, not a liability on the battlefield. Death aversion manifests in the data as group-fights around different, but strongly fortified heroic identities—the ODA's ability to subdue the enemy with incredible tactical precision and the AF SOST's ability to save life with medical precision. Generally, AF SOSTs have come to amicable solutions given time and rapport building with the ODA. This is especially true once the ODA is able to see the life-saving power of immediate damage control surgery on the battlefield. However, if these heroic identities are not relaxed such conflicts could compromise mission effectiveness given even more austere battlefields with higher volumes of casualties as are being anticipated in LSCO.

Additionally, there have been instances where AF SOST medics experience in-fighting in relationship to medical decision making when it pertains to allocating scarce resource. In particular, the challenge has emerged in relationship to either withholding or withdrawing care amidst a mass casualty event. Projecting these same tensions onto a LSCO battlefield suggests that death aversion, could lead to tension and breakdown amidst medical teams if the life-saving identity of medics is not tempered. In other words, mission effectiveness could be compromised from a medical support perspective if triage training does not specifically focus on preparing medics to deal with making life and death choices. Such in-group fighting could be made more difficult to manage in situations of reverse triage, where medics would be forced to let more severely injured casualties wait for medical intervention. When considering the lack of resources, the volume of casualties, and the limited numbers of medical personnel being discussed in LSCO, TMT juxtaposed against our ethnographic data intimates that the singular life-saving focus of combat casualty care culture might engender a culture that occasions ethical breaches if medics are not properly prepared to face death and dying on a multi-domain battlefield.

While such mission erosion and ethical breaches are hypothetical worst-case scenarios based on abstractions about future battlefields, death aversion in the combat casualty care culture has led to more subtle challenges for AF SOST medics who have experienced volumes of death and dying. Next, we will move away from

---

study, the observation that forward surgical teams appear to compete for status power instead of collaborate seems to reinforce the notion that in-group identity causes friction that could decrease surgical interoperability on the battlefield.

extreme challenges that "might be potentiated" by death aversion and discuss the more nuanced burdens that weigh on AF SOST medics who struggle to make meaning of death and dying amidst a redoubled cultural normativity that ignores this reality both in the military and medicine.

12.1.4 *Emergent Themes – Challenges Relative to Death Aversion*

Maintaining a heroic life-saving identity creates unique burdens that impact the medic's sense of self as well as the ability to communally express sorrow and process the loss that has been encountered. We start by describing the how death aversion is woven together with a culture of disenfranchised grief that often leaves medics alone when struggling to find ways to process the depth of loss that has been experienced during deployment. The is doubly true for AF SOST medics who are uniquely positioned at Trauma One medical centers stateside. As such, the volume of traumatic death to which they are exposed is enormous.

When such loss is not culturally acknowledged, medics feel isolated by a military organization that does not value their work or the suffering involved in performing their work. This disenfranchised grief is described in the following narrative:

> When we got home from deployment, we expressed our concern saying, "If this pace continues without time to process it will destroy us as human beings." That was met with some serious pushback, which was difficult. When our pain was ignored, we moved on. It's probably not the best thing to do because at some point we need to face all the skeletons, but we had to move on. Our work doesn't stop when we get home. We immediately continue our military and clinical training schedules. Our schedules are nonstop and so when we get back, so we mostly just bury everything. That means we all carry around a graveyard of casualties we've taken care of who have died. Most of those graveyards are not small. Eventually they get to a level where the death we carry around has more of an impact. At some point, as human beings, we're not able to keep these experiences buried anymore, but there is no way for us to wash out our grief. Everyone is supposed to go through the stages of grief but there is no time and no acknowledgement of the loss. There is a high degree of negativity around my deployment experiences

Some medics described a similar type of disenfranchisement when their sorrow was openly expressed with family members:

> I've seen thousands of people die, but when you have to go through their shit, see a picture of their family with blood all over it, then put a flog on the body, well, I can't make people understand that when I see a flag all I see is the number of flags I've put over people's bodies. At the very beginning of a hockey game the National Anthem was planning. I was standing there looking at the screen and the flag was waving. I just broke down and cried uncontrollably in this arena. My mom and dad were very concerned for me, they didn't know what was going on. I still don't even know how to explain what happened. When I hear the National Anthem and see the flag there's deeper meaning behind it that no one else knows. I've seen the sacrifice that our service members make. I've seen the worst of the worst in what war does to people. When I broke down, I was told I needed to see a counselor to learn to cope with those triggers.

The point of this narrative is not to imply that seeking assistance through mental health is inappropriate. However, it illustrates that normal expressions grief are perplexing and are given meaning by labeling them as "triggers." Such clinical interpretations label grief as pathological, abnormal or something with which the AF SOST medic must "cope." They do not interpret grief as a natural, healthy, and important way to process the experience of death and dying. Similarly, clinical meaning-making leaves medics alone in their grief and does not highlight the value of collectively sharing emotional expression with the military, family, or political communities. In other words, disenfranchised grief separates medics from an ability to authentically express or narrate their deployment experiences without making them pathological or problematic.

During one interview a research team member came up with the notion of ambivalent loss. This culture of disenfranchised grief leads the medic to an enormous sense of ambivalence relative to all the loss experienced as a result of their military and medical service. It is explained by the following narrative:

> Loss is registered but there was nothing we could do about it. We couldn't do anything. Talking about it is not going to make it go away. People try to rationalize it away all day long saying things like "Consider the political objectives X, Y, or Z." Or they try to tell you that you did everything right medically. We will always be able to rationalize the loss of life, but no one is ever going to turn off the sensitive side of our personalities—at least they should not want us too. These experiences come at a very sharp price. No one will be the same after deployment and most people don't get that loss, they try to make it all better or act like it doesn't exist. It all takes a toll and most people do not acknowledge that toll or the alteration to our identity. There is an expectation that people come back the same and just get back to work. I've deployed over a dozen and every time I struggled with ambivalence about the person I had become and about everyone else around me.

Within a culture of disenfranchised grief, medics seem to become at best ambivalent about their loss and struggle to find modes of meaning making or processing that resonate with their lived experience of loss and service.

Ambivalent loss is also exacerbated by the experience of ambiguous loss for AF SOST medics who have experience high number of civilian casualties. Ambivalent loss is a theoretical construct that was articulated by Pauline Boss when researching wives of military pilots taken prisoner in the Vietnam conflict. Boss established that the unknown fate of the husband was a phenomenon different than that of death because the latter creates an opportunity for closure through culturally normative grief processing practices (Boss 2009). In the highly technological world of combat casualty care, American casualties are moved to higher levels of definitive care never to be seen again. In contrast, civilians from the host nation are often transported to less technologically advanced medical facilities as described by the following narrative:

> We do everything we can to fix someone, and in a lot of other countries they just don't have the resources to do that. When we fix someone in a foreign country, we do the best we can at that moment, but then we send them on to their level of care that exists in the hospitals in their environment, and that doesn't necessarily mean that they're going to get equivalent care. Ultimately, we never knew the outcome of the people we worked on. We'd patch them

> up and then the locals would put them in their make shift ambulance, and the patient would disappear. We wouldn't get any real feedback. Did we make a difference in the outcome? When the answer is 'I don't know', that's terrible because we have to live with the uncertainty.

Ambiguous outcomes leave medics with a sense of open-ended interpretation concerning the value of their medical care as well as an ability to create closure. According to Boss, if unaddressed, individuals experiencing ambiguous loss seek closure through dichotomous and concrete answers about life and death (Boss 2009). In other words, they seek to know if what they did was right or wrong. Unfortunately, the desire to find closure where there is none only exacerbates the ambiguity. It also acts as a reinforcement in the quest to affirm or deny the medic's heroic identity. The practical result is the medic moves back and forth between feeling affirmed or denied in that identity. In interviews medics pondered whether unknown cases on which they dwell are also bothered their teammates. In other words, ambiguous loss is an experience that is not readily shared with fellow medics.

When considering the reality many medics are facing in a military and medical cultures that are saturated with death aversion, the challenge is immense. To reiterate, our goal is NOT to suggest that medics or the military medical culture is unethical for focusing on life-saving medical interventions. We walk a tight rope in this paper recognizing that it is important and necessary to research and train the highest quality of life-saving medical interventions, while also suggesting that this remains insufficient if we want to produce the most ethical and effective medics for the future fight. If death aversion is not addressed, a future group of medics that face austere and intense fighting will could crumble against insurmountable odds when trying to achieve an unrealistic 98% survivability. Somehow the normative thrust of this statistic must be altered so that the heroic life-saving identity can be relaxed to slowly accept death as a medical and military reality on the battlefield.

12.1.5 Possible Ways Forward

As mentioned in our introduction, we are touching on deep roots that are strongly grounded in Western medicine and military cultures when discussing death aversion. For that reason, we will not attempt to resolve this situation with strong normative conclusion. In fact, seeking concrete resolution or precise answers to this implicit cultural normativity about life-saving medical interventions might cause more harm than good because both TMT and Ambiguous Loss theory suggest that concrete black and white answers to unknown realities actually reinforces the need to preserve heroic identities. Our goal is not to dismiss or cast aspersions on medics for having a deep desire to save all casualties. However, as we have shown from our ethnographic data with AF SOST medics, death aversion does not come without a cost. This is likely to be amplified on even more austere future battlefields if this cultural phenomenon is not explicitly counteracted. TMT and Ambiguous Loss theory also provide insight into some of our preliminary suggestions. Both theories state that developing a thick relationship with death and outcome ambiguity is critical to relaxing the grip of heroic identities. In thinking about what type of relationship with death might help to allow room for movement within medics' identities, we constructed a second visual diagram that is meant to illustrate a more holistic relationship to life and death that on a graphic plane (Diagram 12.2).

If human existence is to be experienced in its fullness, i.e. in all quadrants of the graphic plane, then death to identity and physiology must be experienced as well as life to identity and physiology. Our goal was to integrate the upper the left upper quadrant as well as the right lower quadrant to include healthy interaction between life and death. If the reader imagines life and death in and asymptotic relationship, then the medic is no longer in a zero-sum game wherein the only good outcome is saving the life of a casualty. The pathological relationship to death would be intentionally aiming to destroy the identity or physiological processes of any casualty, not necessarily choosing to withhold or withdrawal medical interventions.

Practically speaking, we believe that AF SOST medics serve as an exemplar for where to begin to develop a thick relationship with death as well as ways to begin to expand this relationship. As mentioned in the previous section, AF SOST medics prepare at civilian Trauma One Centers. This unique function allows them to have experienced many traumatic deaths before they are deployed to provide medical care in austere environments. We believe that this is a necessary starting point to prepare future medics for multi-domain battlefields in LSCO. However, it is not sufficient. Our ethnographic data unilaterally supports the need for triage training for mass casualties in austere environments—specific training that approximates the complex decision making and wound patterns seen in austere deployment settings.

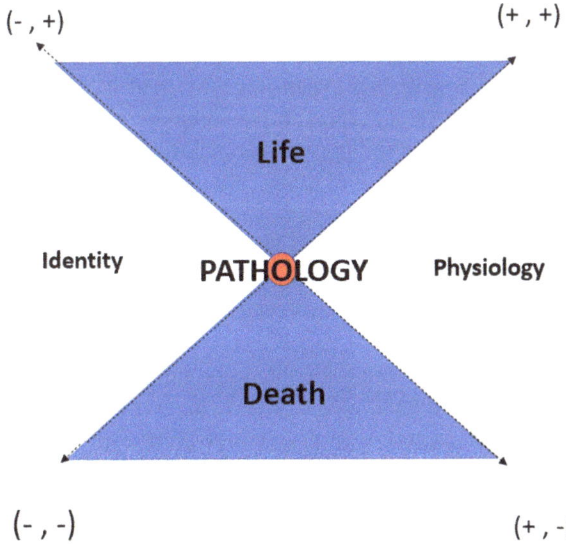

Diagram 12.2 Holistic relationship to life and death. (Constructed by Erika "Ann" Jeschke)

This would potentially increase medics' medical and emotional familiarity with what counts as preventable injuries and survivability in resource depleted environments. One of our secondary goals of research, is to develop AF SOST ethnographic data into more explicit cases for training triage and mass casualty scenarios in austere and unconventional environments.

While we strongly believe and have argued that explicit training directed toward learning military medical decision making in austere mass casualty needs to be researched, such research does not address the harms done to medics from disenfranchised grief. Interestingly, TMT and Ambiguous Loss Theory suggest that developing a rich relationship with life requires engaging in grief processing practices that simple forms of simple meaning-making that fosters heroic idealization of combat experiences. This is also evidenced by some of the narratives from our preliminary data:

> When I came back, I had struggles and I would probably be more abnormal if I didn't. I 100% believe in counseling, but having seen many mental health providers it was somewhat frustrating. There were so many people who said stop count to ten and think about a bunny rabbit. It was totally disheartening. Everyone wanted me to go back to the person I was before I deployed. Finally, I met one mental health provider who understood. That provider said "You're never going be the person you were before. You have to allow yourself to mourn the loss of that person." Counseling has to be geared at optimizing who the service member is now, not focused on some ideal past image of who the service member was before they left for deployment. You can be a better person you can be a stronger person, but you are never going to be ***that*** person. It was like a huge weight was lifted off of me.

Instead of relying on cognitive behavior models of intervention, we suggest that the existential psychology of Viktor Frankl and Rollo May would prove more helpful. These forms of therapeutic intervention are important because of the worldview in which therapeutic interventions are circumscribed. Both Frankl and May believe that death, loss, destruction, ambiguity, and anxiety are natural to the human experience (Frankl 1988; May 1994). More importantly, they are often the point at which the human being finds growth and expansion. The goal is not to create concrete resolution through definitive answers or instantiation of a permanent identity. Instead, it is to learn to live in the grey-zones of life becoming more comfortable with uncertainty and the experience of living into the "maybe." According to Frankl (1985) meaning is made in in the gap between who someone is and who someone wants to be. For May (1994) anxiety and loss of entrenched identity are the birthplace of creativity. Focusing on deployment experiences as fertile grounds for creativity might be amenable to working with medics' whose grief has been overlooked and undervalued.

While these first two suggestions are able to be implemented with relative ease, they are focused on the individual medic. The medical model suggests that healing is focused on the individual and often psychological interventions can tend toward rugged individual bootstrapping. When thinking about the deeper tendrils of the death averse military and medical cultures, we strongly object to placing the burden of healing and recovery on the individual medic. The more challenging aspects of this reality are the cultural roots that run deep and implicitly shape ways of being, expectations, and identities. It is not possible for an individual medic to will-power through a culture and stand against it. While cultural and organizational change management are slower processes, they are essential when thinking about relaxing the burdens of death aversion for the medic in preparation for future battlefields.

When reflecting on death aversion in the military and medical cultures, we have come to suspect that rich narratives are essential to resilient cultures that support medics' effective and ethical decision making. In working through the analysis of this paper, we came to realize that the singular focus on life-saving medical interventions forecloses any opportunity to develop narratives around the "good" combat death because death is always a bad outcome. In future project we intend to ponder how narratives can begin to shape new cultures around good medicine, the good medic and a good combat death. Our goal is to eventually soften cultural barriers that wrap themselves around medics and make it difficult for them to integrate and heal from their deployment experience. Finally, our long-term goal is to develop modes of palliative performance with our ethnographic data that can help the political community better understand the experiences of military medics as well as help reduce the tumultuous experience of returning to the civilian community. While much of our work is conjectural, we hope that our small efforts help to reframe the reality of military medicine and support an ethical, medical, and military reality of death in combat.

References

Becker, E. 1997. *The denial of death*. Simon and Schuster.

Boss, P. 2009. *Ambiguous loss: Learning to live with unresolved grief*. Harvard University Press.

Butler, F.K., D.J. Smith, and R.H. Carmona. 2015. Implementing and preserving the advances in combat casualty care from Iraq and Afghanistan throughout the US military. *Journal of Trauma and Acute Care Surgery* 79 (2): 321–326.

Davis, M.R., and T.E. Rasmussen. 2019. Winds of change in military medicine and combat casualty care. *Journal of Trauma and Acute Care Surgery 87* (1S): S1–S2.

Ehlers, N., and S. Krupar. 2019. *Deadly biocultures: The ethics of life-making*. University of Minnesota Press.

Frankl, V.E. 1985. Logos, paradox, and the search for meaning. *Cognition and psychotherapy*, 259–275.

Frankl, V.E. 1988. *The will to meaning: Foundations and applications of logotherapy*. Expanded ed. Penguin.

Greenberg, J., and A. Arndt. 2011. Terror management theory. In *Handbook of theories of social psychology: Collection: Volumes 1 & 2*, ed. P.A.M.V. Lange, A.W. Kruglanski, and E.T. Higgins, 398–415. SAGE.

Janak, J., Z. Stockinger, E. Mazuchowski, R. Kotwal, J. Sosnov, H. Montgomery, F. Butler, S. Shackelford, J. Gurney, M. Spott, L. Finelli, and D.J. Smith. 2018. Military preventable death conceptual framework: A systematic approach for reducing battlefield mortality. *Military Medicine* 183 (suppl_2): 15–23. https://doi.org/10.1093/milmed/usy149.

Jeschke, E. 2019. Damage control Interdisciplinarity: An antidote to death despair in military medicine. *Louis UJ Health L. & Pol'y* 13: 201.

Kaempf, S. 2018. *Saving soldiers or civilians?: Casualty-aversion versus civilian protection in asymmetric conflicts*. Cambridge University Press.
Kübler-Ross, E. 1969. *On death and dying*. Scribner.
Lacquement, R.A. 2004. The casualty-aversion myth. *Naval War College Review* 57 (1): 38–58.
May, R. 1994. *The courage to create*. WW Norton & Company.
National Academies of Sciences, & Medicine. 2016. *A national trauma care system: Integrating military and civilian trauma systems to achieve zero preventable deaths after injury*. National Academies Press.
Solomon, S., J. Greenberg, and T.A. Pyszczynski. 2015. *The worm at the core: On the role of death in life*. Random House Incorporated.

Erika "Ann" Jeschke is currently working with the Air Force Research Lab, the Army Institute for Surgical Research, and Navy Medical Research unit on a program of research to develop a broader understanding of the impact of catastrophic injury exposure on various forward surgical trauma teams in the United States Military. Dr. Erika "Ann" Jeschke, PhD is currently working with the Air Force Research Lab, the Army Institute for Surgical Research, and Navy Medical Research unit on a program of research to develop a broader understanding of the impact of catastrophic injury exposure on various forward surgical trauma teams in the United States Military. The goal of this research is to develop a theory of performance for trauma clinicians that is framed by culturally sensitive and salient attributes of well-being. Dr. Jeschke is also a Senior Fellow at the University of Chicago's MacLean Center for Clinical Medical Ethics. Her interdisciplinary research is grounded in the following question: What does it mean to care amidst violence? Relying on medical humanities and anthropology, she seeks to expand the medical context of healing beyond the clinical context and physician-patient dyad to include communal rituals as a means of enhancing resilience. In this work, Dr. Jeschke defines resilience as dynamic freedom of maneuver that is sustained and enhanced through positive grief processing practices. She is working to uncover how embodied practices of caring are the central attribute of character development and performance in military and civilian trauma teams.

Dr. Hannah R. Martinez is a licensed clinical psychologist currently assigned to the 11th Airborne Division, 2nd Brigade Combat Team (Airborne) as the Senior Behavioral Health Officer. Dr. Martinez graduated with her PhD in clinical and military psychology from the Uniformed Services University in 2021. She conducted a mixed-methods dissertation study looking at the impact of romantic relationships on suicidal ideation and attempt in an active duty military population. She has also served as co-investigator on projects examining the phenomenology of suicide, suicide prevention in the military, and the formation of resiliency in a highly specialized military population. Dr. Martinez has presented research at various national conferences including the American Psychological Association, the Association for Behavioral and Cognitive Therapies, the Military Health Systems and Research Symposium, the American Association for Suicidology, and the International Society for Military Ethics.

Eleanor M. Choi is an experienced researcher with degrees in theology and psychology from Georgetown University. After graduating, Ms. Choi spent a year working for the Suicide Care, Prevention, and Research Initiative where she contributed to several projects investigating suicide in military personnel and their dependents. Ms. Choi is currently a senior research analyst with Advisory Board, a best-practice research organization serving health care entities including providers, payers, and pharmaceutical manufacturers, among other industry members.

Colonel John R. Dorsch, DO (ret) enjoyed a distinguished career in the United States Air Force as a leader in operational medicine where he finished as the senior medical officer for Special Tactics and Pararescue medical director. He was a Chief Flight Surgeon with numerous deployments over five different continents logging 470 combat flight hours in support of special operations, combat search and rescue, casualty evacuation, and mobile surgical and resuscitation teams. Colonel Dorsch currently serves as a staff emergency physician at UNC Southeastern Medical Center, clinical faculty at Campbell University, operational team member on Vigilint's Mobile Surgical Resuscitation Team, and Assistant Medical Director to Robeson County Emergency Medical Services. Colonel Dorsch continues to drive advances in military operational medicine as Senior Medical Advisor to SimX, the world's leader in virtual reality medical simulation, in addition to providing consultation on several DoD sponsored R&D efforts in trauma and austere medicine.

Lt Col Sarah L. Huffman is currently assigned to Travis Air Force Base, CA as the Director of Biobehavioral Research in the Clinical Investigation Facility (CIF). Lt Col Huffman was previously assigned to Wright Patterson Air Force Base in OH as a nurse scientist in the Airman Biosciences Division, En Route Care Research/711th Human Performance Wing, and as Director of the Clinical Investigations Facility at the 88th Medical Group. Lt Col Huffman earned a Bachelor of Science Degree in nursing from the University of Texas, a master's degree with a focus on critical care nursing from the University of Virginia with certification as a critical care Clinical Nurse Specialist and Acute Care Nurse Practitioner, and a PhD in nursing.

GPSR Compliance
The European Union's (EU) General Product Safety Regulation (GPSR) is a set of rules that requires consumer products to be safe and our obligations to ensure this.

If you have any concerns about our products, you can contact us on

ProductSafety@springernature.com

In case Publisher is established outside the EU, the EU authorized representative is:

Springer Nature Customer Service Center GmbH
Europaplatz 3
69115 Heidelberg, Germany

www.ingramcontent.com/pod-product-compliance
Ingram Content Group UK Ltd.
Pitfield, Milton Keynes, MK11 3LW, UK
UKHW022211051026
15624UKWH00001B/106
9783031290589